U0144737

美國綠建築LEED
基礎知識與應用

Green Associate認證 第二版

江軍、林巧文 著

五南圖書出版公司 印行

推薦序一

　　當你開始閱讀這本書時，恭喜你！因為這是作者江軍先生針對美國綠建築協會 LEED 的評級系統，做正確完整、深入淺出的介紹與解析。同時你也可從這本書中獲得明確且豐富的資訊，協助你了解 LEED 每一項評估類別在制訂時的真諦，和每一個策略對環境所造成的衝擊。江軍先生一本以往專精研磨求知的態度，集聚長期對永續設計的熱忱，並累積留美期間實地勘察運作的經驗，充分掌握綠建築的特色以及其應用的功能，從他專業學者的角度，精心地撰寫解說與論述，實在是一本難能可貴的著作。

　　近幾年來，世界氣候變遷的快速，能源缺乏供應的起伏，人口指數倍級的暴增，農糧生產損失減量，提升了廣泛大眾對永續發展的需求與認知。同時醞釀出無法過止的動力，使綠建築成為必要的建設條件。身為美國永續建築資深學者的我，非常佩服作者將 LEED 的每一個類別自概念到優惠成果，以非常簡易近人的文筆，有條有理地闡述清晰，並提供相關的模擬試題。參閱本書將會使讀者具備十足的信心，並且輕易地通過驗證的考試。儼然這是一本永續建築設計師不可或缺的參考書籍。

吳和甫（Hofu, Wu）博士
美國建築學院資深院士
於美國洛杉磯鄭重推薦

推薦序二

　　認識江軍是在我數年前的 LEED GA 課程上，江軍本身是台大土木系高材生，對於綠色永續建築有著濃厚的興趣，課程期間將 LEED 綠建築評估系統學習得很透徹，並順利通過 LEED GA 考試，之後也時常回來擔任我們 LEED 課程讀書會的小老師，與學員分享 LEED 知識與 GA 考試經驗，並協助學員們掌握考試重點。巧文是留美歸國的 LEED AP，曾經是我綠建築顧問公司的優秀同仁，執行過不少 LEED 綠建築專案，也是我們 LEED GA 課程的主要講師，LEED 執行與教學經驗相當豐富。

　　台灣一直沒有出版過 LEED 相關繁體中文版書籍，很高興得知江軍與巧文合作，將多年來累積的 LEED 知識與應試心得，轉化成這本繁體中文版 LEED GA 參考書籍。本書的內容架構詳盡，囊括綠建築的基本概念、LEED 認證評估系統內容以及準備 LEED GA 考試所需要的相關資訊。國人學習 LEED 一直以來最大的障礙，就在對於原文資料相關專有名詞與英文用語字義的解讀，因此在學習過程中常覺得深奧難懂，或產生挫折感。翻譯 LEED 綠建築評估系統，除了需要具備優秀的英文能力之外，仍需了解綠建築相關英文專有名詞，才能翻譯到位，這本書在兩位作者的努力下，在專業度與翻譯精準度都相當到位，讓不諳英文的人也能輕易讀懂 LEED，降低學習上的門檻。

有興趣想了解 LEED 綠建築評估系統的朋友，本書是本值得推薦的
入門書，要準備 LEED GA 考試的朋友，更是不能錯過！

美國綠建築協會院士（USGBC LEED Fellow）
台灣綠領協會（TGCA）理事長
SSDC 澄毓綠建築設計顧問公司總經理
Enertek 川昱永續環控有限公司執行長

作者序一

　　本書是全台灣第一本專門針對美國綠建築評估體系介紹的中文教材，隨著全球氣候變遷與環保意識的抬頭，越來越多人重視綠建築的設計理念。在台灣，除了本土性的 EEWH 評估系統外，也相當多的建築物希望取得 LEED 認證為其加分，以增加其國際的市場。更因如此，越來越多的建築相關從業人員朝著 LEED 專業認證人員邁進，而 LEED 認證考試雖然已於 2014 年前開放了中文輔助語言，但是由於缺乏相關教材與資料，也讓很多有心想認識或準備 LEED 考試的夥伴有了閱讀語言的障礙，也因此我與巧文老師開始將資料整理成中英對照的模式，以方便讀者學習。

　　第一本書的出版，心中的感覺非常特別，有如懷胎十月後將心血整理出來的感受。首先非常感謝吳和甫教授，讓我接觸並開啟了綠建築大門，一窺這豐富有趣的領域。接著感謝綠領的陳重仁理事長，以及諸位 LEED 課程的講師，你們的經驗傳承我確實收穫良多，以及巧文老師的幫助與資源，才能讓本書順利付梓。也感謝五南出版社的王正華主編與金明芬編輯認真且非常無私的幫助，讓出版過程非常的順利。最後要謝謝支持我的父母與朋友，因有你們的支持讓我可以東奔西闖的去學習我感興趣的事情，也才讓今天的我有與同齡者不同的經歷。

　　最後希望讀者可以使用本書於 LEED 系統的學習，或是用來針對 LEED 認證考試當作學習的材料。本書附錄有詳盡的考試步驟說明與中

英文專有名詞對照表，相信對要考試的讀者有相當大的幫助。末了祝各位學習順利，並順利取得 LEED 認證人員，一起爲永續的環境設計盡一份心力！

　　本書撰寫過程雖力求盡善盡美，仍有疏漏不足之處，敬請來信指教（gem6004@gmall.com），非常感謝您寶貴的意見。

江軍　敬識

作者序二

從 2007 年第一次在美國舊金山藝術大學一堂選修課程接觸到綠建築設計的時候，我才發現，氣候變遷、地球暖化等等的環保議題其實離我們並不遙遠，身為設計師／建築師的我們，在每一個下筆畫圖的瞬間，都有機會可以為我們的地球帶來一點點的改變。當時的我默默地在心中許下了一個願望，希望自己能夠把這樣一個好的概念帶回台灣，當一個推廣者，讓更多的人認識綠建築的優點。

很幸運的，因為在美國取得了 LEED AP 證照，帶領我回台灣順利地找到了第一份工作—綠建築設計顧問，有幸為國內最頂尖一流的高科技公司提供綠建築的設計顧問服務。2013 年 3 月份更是夢想成真的時刻，全台灣第一堂的 LEED GA 認證輔導課程終於在公司同仁的共同努力下，成功在文化大學推廣教育部開跑。我也成為課堂講師群之一，帶領著眾多有熱情、有遠見的學員一同來了解綠建築的概念，並認識 LEED 認證系統成為全球最廣泛被運用的各種原因，一路至今已累積協助超過 200 名以上的學員順利取得證照。

作者江軍先生也是當年眾多的學員之一，他本著無人能及的學習熱誠與效率，為班上同學主動提供各種整理的筆記及心得分享，幫助後面無數學員能更輕鬆順利的考取證照。當他第一次跟我提起準備要出書，我一度擔心的認為這會是條辛苦的道路，不過積極、努力的他，總是一次次過關斬將，讓人感受到年輕人該有的韌性與活力。

這一本書的誕生，意味著可以幫助更多的人接觸，並深入了解這個源起於美國的 LEED 綠建築認證系統，如何透過系統化以及量化的概念，鼓勵設計單位能夠輕鬆的達到對環境友善的建築設計。在此感謝忙碌之餘，仍然行動力不減的江軍能夠代替我來實現這樣的夢想，謝謝！

最後送一句話給正在讀這本書的你，You must be the change to the world you wish to see! 希望大家都一起來為我們的地球盡一份微小的心力，哪怕只是一點點，也都是一個改變的起點！

林巧文　敬識

目錄 (Table of Contents)

第一章　綠建築永續設計基本概念（Basic Concepts of Sustainable Green Building Design）

學習目標

- 什麼是綠色建築？（What is a green building?）
- 什麼是高效的綠色建築物？（What makes a building highly efficient and green?）
- 認識各種綠色建築評級系統（Understand various green building rating systems）
- 認識 LEED 的意義（Understand the significance of LEED）
- 各種 LEED 評估系統（Learn about the different LEED rating systems）

1.1 綠建築簡介（Introduction to Green Building）

　　綠色建築或綠建築（Green Building）係指本身及其使用過程在生命週期中，例如選址、設計、建設、營運、維護、翻新、拆除等各個建築階段皆達成環境友善與資源有效運用的一種建築。也就是說，綠色建築於設計上試圖在人造建築與自然環境之間取得一個平衡點。這需要設計團隊、建築師、工程師以及客戶在專案的各階段中緊密合作。而綠色建築的實作發展補足傳統建築設計對於經濟效益、實用性、耐用度以及舒適度方面的不足。（Green Building refers to a structure and its processes that are environmentally friendly and resource-efficient throughout its entire lifecycle, including stages such as site selection, design, construction, operation, maintenance, renovation, and demolition. In other words, green building aims to achieve a balance between man-made structures and the natural environment in its design.）

　　綠建築相近觀念爲自然建築（Natural Building），是指正常用於較小的比例，並偏向使用當地可取用的材料。其他常見的同義詞包括了永續設計（Sustainable Design）及綠色建築物（Green Architecture）。（A similar concept to green building is Natural Building, which typically applies to smaller structures and emphasizes the use of locally available materials. Other common synonyms include Sustainable Design and Green Architecture.）

　　今天環境面臨的挑戰包括了（Today, the environment faces several critical challenges, including:）

- 氣候變遷（Climate change）
- 資源枯竭（Resource depletion）
- 臭氧層破洞（Ozone layer depletion）
- 土地汙染（Land pollution）
- 水汙染（Water pollution）
- 空氣汙染（Air pollution）

　　建築物直接製造了上述對環境的汙染。而綠色建築是面對這些環境問題時，可以提供解決的方案。（Buildings directly contribute to the pollution of the environment in these areas. Green building offers solutions to address these environmental challenges.）

　　下圖顯示那些排放 CO_2 較高的海灣合作委員會（GCC）國家。這可能是因爲國家的氣候條件、生活風格，與缺乏基礎設施有關的公共交通工具等原因而造成高碳排放。（The chart below shows the Gulf Cooperation Council (GCC) countries with higher CO_2 emissions. This is likely due to factors such as climate conditions, lifestyle, and a lack of public transportation infrastructure, which contribute to high carbon emissions.）

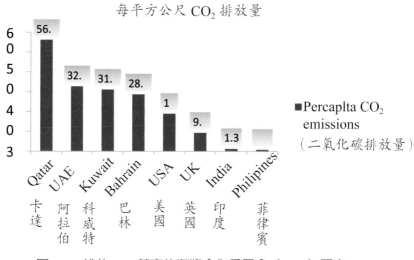

圖 1-1　排放 CO_2 較高的海灣合作委員會（GCC）國家

下圖顯示，綠色建築物可以顯著減少資源在建築中的使用量，以及減少對於環境的影響。（The diagram below shows that green buildings can significantly reduce resource usage in construction and minimize their impact on the environment.）

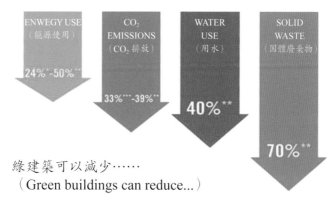

圖 1-2　綠建築可減少的資源使用量

1.2 綠建築定義（Definition of Green Building）

什麼是綠色建築？（What is a green building?）

　　簡單的定義就是將永續的概念導入，以全面整合性的方式來改變傳統的建築設計、施工以及使用的模式。（Simply defined, it is the integration of sustainable concepts into the design, construction, and usage of buildings in a holistic way, transforming traditional methods.）

綠色建築是……（A green building is characterized by:）

- 高效率的能源使用（Efficient energy use）
- 高效率的用水（Efficient water use）
- 提供更好的室內環境及因此更好的生活條件（Providing better indoor environments and, as a result, better living conditions）
- 使用環境友好或可持續利用的建築材料（Using environmentally friendly or sustainable building materials）
- 減少生產浪費（Reducing production waste）
- 有較少的交通運輸需求（Lower transportation demands）
- 保護／恢復生物棲息地（Protecting or restoring natural habitats）

什麼是永續發展？（What is Sustainable Development?）

　　世界環境發展委員會（WECD）對永續發展下了這樣的定義：（The World Commission on Environment and Development (WCED) defines sustainable development as:）

　　「永續發展是合理的滿足當代的需求，但不危及後代滿足他們需求的能力。」（"Sustainable development is meeting the needs of the present without compromising the ability of future generations to meet their own needs."）

圖 1-3　永續發展解決方案

　　因此，永續發展包含了經濟、人文和環境三個面向。一個好的發展必須兼顧此三個面向，才能達到合理的平衡。（Thus, sustainable development encompasses three dimensions: economic, social, and environmental. A successful development must balance these three aspects to achieve a reasonable equilibrium.）

1.3 為什麼需要綠色建築？（Why Do We Need Green Buildings?）

- 土地的開發往往會破壞野生動物的棲息地。（Land development often destroys wildlife habitats.）
- 開採、生產與運輸建築材料的過程中，容易造成水和空氣汙染，釋放有毒化學物質，並且排放溫室氣體。（The extraction, production, and transportation of building materials can easily cause water and air pollution, release toxic chemicals, and emit greenhouse gases.）
- 建築施工、營運需要大量消耗能源和水，並產生大量的廢物流。（Building construction and operation consume large amounts of energy

and water and generate significant waste.）

- 綠建築會在整個建築生命週期持續減少對環境的影響，建築的生命週期與成本降低數據如下（根據統計資料）：（Green buildings continuously reduce environmental impact throughout the building's entire lifecycle. The lifecycle cost reduction data (based on statistics) is as follows:）

  ➢ 建築物平均節能 24~50%（Buildings save an average of 24-50% in energy）

  ➢ 建築物碳排放平均降低 33~39%（Carbon emissions are reduced by an average of 33-39%）

  ➢ 建築物平均節水 40%（Buildings save an average of 40% in water usage）

  ➢ 降低固體廢棄物達 70%（Solid waste is reduced by up to 70%）

1. 綠建築的優點（Benefits of Green Buildings）

- 降低建築物營運費用（Reduces building operational costs）
- 縮短投資回收年限（Shortens the investment payback period）
- 增加建築物價值（Increases the value of the building）
- 增進能源與用水效率（Enhances energy and water efficiency）
- 減少廢棄物量（Reduces waste）
- 創造健康舒適的生活環境（Creates a healthy and comfortable living environment）
- 降低溫室氣體排放（Lowers greenhouse gas emissions）
- 提升企業形象（Improves corporate image）

2. 建築物能源使用消耗量（Energy Consumption in Buildings）

下圖的能源分布統計是由美國統計而來，所以我們可以發現因為位

於溫帶地區，建築物開暖氣的情況比冷氣還多，而同理如果建築物位於亞熱帶或熱帶地區，能源的使用最重則會變成冷氣。（The energy distribution statistics shown in the chart are based on data from the United States. We can observe that, due to the temperate climate, heating is used more frequently than air conditioning. Similarly, if a building is located in subtropical or tropical regions, the largest energy consumption would be for air conditioning.）

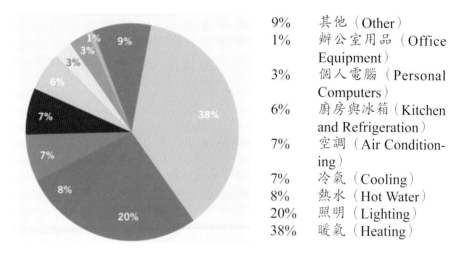

9%	其他（Other）
1%	辦公室用品（Office Equipment）
3%	個人電腦（Personal Computers）
6%	廚房與冰箱（Kitchen and Refrigeration）
7%	空調（Air Conditioning）
7%	冷氣（Cooling）
8%	熱水（Hot Water）
20%	照明（Lighting）
38%	暖氣（Heating）

圖 1-4　建築物能源使用消耗量

3. 綠建築的必要性？（The Necessity of Green Buildings）

使用交通運輸對整體環境的影響層面非常廣泛，其中包含交通系統大量消耗燃料、產生噪音、二氧化碳及其他空氣汙染等，對環境造成許多傷害。所以，一個好的綠建築不只是單單指建築物本身，而是包含整體周圍的開發也維持一個永續綠色交通的方式。（The impact of transportation on the overall environment is extensive, including the large consumption of fuel by transportation systems, noise generation, carbon dioxide emissions, and other air pollutants, all of which cause significant harm to the environment. Therefore, a good green building is not just about the building itself but also involves maintaining sustainable and green transportation methods in the surrounding development.）

1.4 綠色建築的核心原則（Core Principles of Green Building）

可以從以下六個 LEED 核心單元來說明 LEED 的綠建築架構以及每個章節所包含的重點：（The LEED green building framework and the key points of each section can be explained through the following six core LEED categories:）

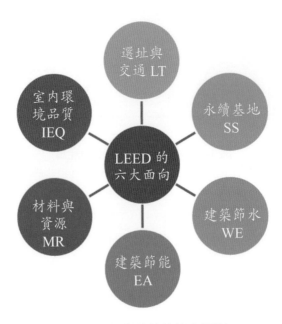

圖 1-5　綠色建築的核心原則

1. 選址與交通（Location & Transportation）

- 基地區位（Site location）
- 綠色交通（Green transportation）

2. 基地永續性（Site Sustainability）

- 生態環境（Ecological environment）
- 基地綠化與開放空間（Site greening and open spaces）
- 基地保水（Stormwater management）
- 熱島效應降低（Reducing heat island effect）
- 夜間光害降低（Reducing light pollution）

3. 建築節水（Water Efficiency）

- 植栽澆灌節水（Water-efficient irrigation for landscaping）
- 用水器具節水（Water-efficient fixtures）
- 雨水回收再利用（Rainwater harvesting and reuse）
- 中水回收再利用（Greywater recycling）
- 生活汙水減量與再利用（Reducing and reusing wastewater）

4. 建築節能（Energy Efficiency）

- 建築外殼節能（Energy-efficient building envelope）
- 空調系統節能（Energy-efficient HVAC systems）
- 照明系統節能（Energy-efficient lighting systems）
- 熱水系統節能（Energy-efficient hot water systems）
- 動力系統節能（Energy-efficient power systems）
- 可再生能源系統：太陽能、風能、重力能等（Renewable energy systems: solar, wind, gravitational energy, etc.）
- 功能驗證（Commissioning）
- 能源監控與管理（Energy monitoring and management）

5. 材料與資源（Materials & Resources）

- 營建廢棄物管理（Construction waste management）
- 建築生命週期環境衝擊評估（Environmental impact assessment of building lifecycle）
- 既有建築再利用（Reuse of Existing Buildings）
- 環保永續建材（Environmentally sustainable materials）
- 建材分析和優化（Material analysis and optimization）

6. 室內環境品質（Indoor Environmental Quality）

- 室內空氣品質（Indoor air quality）
- 室內空氣汙染源控制（Control of indoor air pollution sources）
- 低逸散性健康綠建材（Low-emission, health-friendly green building materials）
- 舒適溫熱環境（Thermal comfort）
- 自然採光利用（Use of natural daylight）
- 聲環境表現（Acoustic performance）

(1) 綠建築最重視（Key Focus of Green Building）

- 建成環境的生命週期（Life Cycle of Built Environments）
- 整合式的手法（Integrative Approach）
- 綠建築的成本與效益（Green Building Costs and Benefits）

(2) 建築生命週期（Building Life Cycle）

- 建築材料從哪裡來？ 使用完畢後都去了哪裡？（Where do the building materials come from? Where do they go after use?）
- 能源與水資源為何？ 使用這些資源對建築物周遭環境有何影響？（What are the sources of energy and water? What is the impact of using these resources on the surrounding environment?）

- 建築物使用者將會使用何種交通工具？（What transportation methods will building occupants use?）

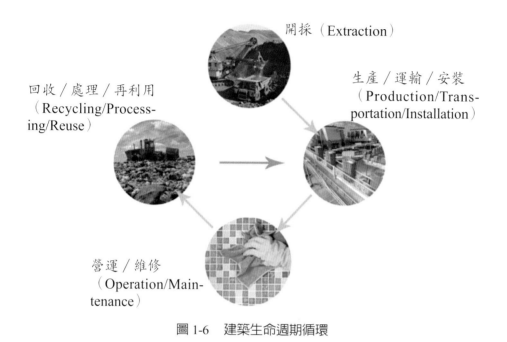

開採（Extraction）

生產／運輸／安裝（Production/Transportation/Installation）

回收／處理／再利用（Recycling/Processing/Reuse）

營運／維修（Operation/Maintenance）

圖 1-6　建築生命週期循環

1.5 綠建築整合式設計流程（Green Building Integrated Design Process）

　　綠建築整合式設計流程是一種協作且全面的方式，從專案的初期階段開始，建築師、工程師、承包商和客戶等所有相關利害關係者都共同參與。透過提早規劃，這個流程能確保永續目標融入建築設計和施工的每一個層面。這種方式不僅提升了能源效率、節水效果及材料利用，還能減少長期營運成本和對環境的影響。越早開始這個流程，越能有效地解決潛在問題，從而達到更優化和永續的建築成果。（The Green Building

Integrated Design Process involves a collaborative and holistic approach, where all stakeholders, including architects, engineers, contractors, and clients, work together from the very beginning of the project. By engaging in early planning, the process ensures that sustainability goals are incorporated into every aspect of the building's design and construction. This approach not only improves energy efficiency, water conservation, and material use but also reduces long-term operational costs and environmental impact. The earlier this process is initiated, the more effectively potential issues can be addressed, leading to a more optimized and sustainable building outcome.）

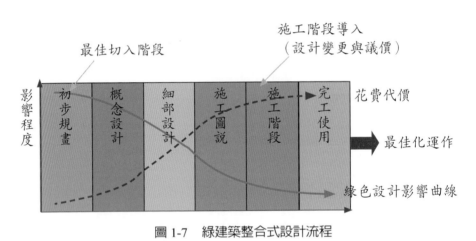

圖 1-7　綠建築整合式設計流程

LEED 綠建築專案執行流程（LEED Green Building Project Implementation Process）

設計階段（Design Phase）
- 專案註冊
- 整理文件
- 圖說準備

發包階段（Bidding Phase）
- 工程分項
- 介面整合

施工階段（Construction Phase）
- 檢驗設計項目
- 功能驗證

認證階段（Certification Phase）
- 取得認證

1.6 綠色建築評估系統（Green Building Rating Systems）

如何建立一個能夠被認證為綠色的建築？（How can a building be certified as green?）

　　建築有各種綠色建築評級系統來認證為綠色建築。綠色建築評級系統是利用工具評估建築物，像能源效率的各個方面，提高用水效率、使用的材料、室內環境品質、基地的位置，如果符合其預設的標準，就給予證書或標章證明。（There are various green building rating systems used to certify buildings as green. Green building rating systems use tools to assess different aspects of a building, such as energy efficiency, water conservation, material use, indoor environmental quality, and site location. If the building meets the pre-set standards, it is awarded a certificate or label as proof of its green status.）

　　在世界各地有眾多的綠色建築評級系統。下面是各種綠色建築評級系統的部分清單：（There are many green building rating systems used worldwide. Below is a partial list of various green building rating systems:）

- 澳大利亞：Nabers / Green Star
- 巴西：AQUA / LEED Brasil
- 加拿大：LEED Canada / Green Globes
- 中國：綠色建築評價標準（GB/T 50378—2006）
- 芬蘭：Promis E
- 法國：HQE
- 德國：DGNB
- 香港：HKBEAM
- 印度：GRIHA (national green rating) / LEED India

- ◼◻ 義大利：Protocollo Itaca
- ◼◻ 墨西哥：LEED Mexico
- ▬▬ 荷蘭：BREEAM Netherlands
- ▦ 紐西蘭：Green Star NZ
- ◉ 葡萄牙：Lider A
- ◼◻ 新加坡：Green Mar kand Construction Quality Assessment System (CONQUAS â)
- ▷ 南非：Green Star SA
- ▬ 西班牙：VERDE
- ▤ 美國：LEED/Green Globes
- ✠ 英國：BREEAM
- ◼ 台灣：EEWH
- ● 日本：CASBEE

以下針對幾項相關的綠建築指標稍微介紹：

中國綠色建築評價標準（China Green Building Evaluation Standard）

目前中國的綠色建築評價標準是 GB/T 50378-2019，該標準於 2019 年實施，並在 2023 年提出了修訂草案。這一標準對綠色建築的評價進行了全面擴充，特別關注安全性、健康性和宜居性，同時也強調傳統的「四節一環」（節能、節水、節地、節材和環境保護）。標準根據以下幾個方面對建築物進行評估：安全與耐久性健康與舒適居住便利性資源節約環境適宜性。（The current China Green Building Evaluation Standard is GB/T 50378-2019, which was implemented in 2019, with draft revisions proposed in 2023. This standard has expanded the scope of green building assessments, focusing not only on environmental protection but also on safety, health, and livability. In addition to the traditional "four savings" (energy, water, land, and materials), it emphasizes:Safety and Durability、Health and Comfort、Convenience for Occupants Resource 、Conservation、Environmental Friendliness。）

該標準的一些特點包括：擴展評估範疇：不僅關注建築物的環保性，還納入了對建築物的安全、健康及可居住性的評價。（Key features include:Expanded Assessment Scope: The standard evaluates the environmental friendliness of buildings, including safety, health, and livability.）

鼓勵創新：包括加分項，鼓勵綠色技術的創新與應用。（Encouragement of Innovation: Additional points are awarded for innovations in green technology.）

碳排放要求：2023 年草案修訂強化了對碳排放強度的報告，特別針對綠色建築認證中的一星及以上評級的專案。（Carbon Emission Requirements: The 2023 draft revisions introduce more stringent carbon emission reporting, especially for projects aiming for a 1-star or higher rating in green certification.）

英國 BREEAM 標準（UK Green Building Evaluation Standard (BREEAM)）

英國綠色建築評估標準（BREEAM）創立於 1990 年，是全球首個綠色建築評估標準，旨在不斷改善人居環境，做到建築開發、環境友好、社會和諧以及經濟可持續發展。（The UK Green Building Evaluation Standard (BREEAM), established in 1990, is the world's first green building assessment standard. Its aim is to continuously improve the built environment by ensuring that building development is environmentally friendly, socially harmonious, and economically sustainable.）

　　1. 能系統地把握 BREEAM 認證的各個技術要點的內涵、實施策略，更加深刻地理解綠色建築的本質。獲得英國建築研究院所（BRE）頒發的英國綠色建築（BREEAM）國際註冊評估師職業資格，有權代表 BRE 對綠色建築專案進行評估並出具評估報告。（Systematically grasp the technical points and implementation strategies of BREEAM certification to gain a deeper understanding of the essence of green buildings. By obtaining the international assessor qualification issued by the UK Building Research Establishment (BRE), professionals are authorized to represent BRE in assessing green building projects and issuing evaluation reports.）

　　2. 基於全生命週期理論，對建築對環境影響的深入調查，這是 BREEAM 評價方法最大的優勢所在，也是對「可持續發展原則」的最大體現。（Based on the whole lifecycle theory, BREEAM conducts an in-depth investigation of a building's environmental impact. This is one of the greatest strengths of the BREEAM assessment method and a significant reflection of the "sustainable development principles."）

　　3. BREEAM 是一種條款式的評價系統，是建築在各個環境表現類別內的「相對綠色」程度，反映了綠色建築在一定的社會技術環境下的相對表現，而不是絕對的「綠色」程度。當新的技術或革新出現而需要增加新的評估條款時，BREEAM 無需改變評價架構、過程及評估結果，可以保持評估的一貫性。（BREEAM is a clause-based evaluation system that assesses the "relative greenness" of a building within various environmental performance categories. It reflects the relative performance of green buildings within a given socio-technical environment rather than an absolute level of "greenness." When new technologies or innovations emerge that require additional evaluation cri-

teria, BREEAM does not need to change its evaluation framework, process, or results, ensuring consistency in assessments.）

　　4. 為了方便建築師在設計階段考慮各設計方案的環境影響，BRE 推出了一種自助式定量建築環境影響評價軟體。它建立了一個龐大的數據庫，提供各種建築元素的環境影響數據。這樣建築師可以在早期設計階段對專案進行環境影響分析。（To help architects consider the environmental impact of different design options during the design phase, BRE introduced a self-assessment quantitative building environmental impact evaluation software. This software has a large database that provides environmental impact data for various building elements. This allows architects to conduct environmental impact analysis in the early stages of project design.）

台灣綠建築九大指標（EEWH）（Taiwan's Nine Indicators for Green Buildings (EEWH)）

　　我國的綠建築係以台灣亞熱帶高溫高濕氣候特性，掌握國內建築物對生態（Ecology）、節能（Energy Saving）、減廢（Waste Reduction）、健康（Health）之需求，訂定我國的評估系統及標章制度，並自 1999 年 9 月開始實施。評估系統有「綠化量」、「基地保水」、「水資源」、「日常節能」、「二氧化碳減量」、「廢棄物減量」、「汙水垃圾改善」、「生物多樣性」及「室內環境」九項指標。（Taiwan's green building evaluation system, known as EEWH, is based on the country's subtropical climate characteristics of high temperature and humidity. It focuses on the ecological, energy-saving, waste-reducing, and health needs of buildings in Taiwan. The evaluation system and certification standards were established and implemented in September 1999. The EEWH system includes nine indicators for evaluating green buildings: Greening

QuantitySite Water RetentionWater ResourcesDaily Energy SavingCO2 Reduc-
tionWaste ReductionSewage and Garbage ImprovementBiodiversityIndoor En-
vironment）

　　2007 年又增訂完成「綠建築分級評估制度」，其綠建築等級由合格至
最優等依序為合格級、銅級、銀級、黃金級、鑽石級等五級，而本分級評
估制度除與國際趨勢同步，也是提升綠建築水準的有效策略，同時我國的
「綠建築」可重新定義為「生態、節能、減廢、健康的建築物」。因此，
台灣的綠建築系統也稱為 EEWH 綠建築評估系統。（In 2007, Taiwan intro-
duced the Green Building Grading System, which ranks green buildings into five
levels: Certified, Bronze, Silver, Gold, and Diamond. This grading system aligns
with international trends and serves as an effective strategy for improving green
building standards. Taiwan's green buildings are now redefined as buildings that
are ecological, energy-saving, waste-reducing, and healthy, making EEWH an
important national system for green building assessment.）

1.7 LEED 優點與趨勢分析（LEED Advantages and Trend Analysis）

為什麼採用 LEED ？（Why adopt LEED?）

　　當有各種綠色建築評級系統，有人會想，為什麼應該要採用 LEED
呢？以下列出幾點原因給大家參考：（When there are various green building
rating systems, one might wonder why they should choose LEED. Here
are a few reasons for consideration:）

- LEED 為國際公認的綠色建築評級系統。（LEED is an internation-
  ally recognized green building rating system.）
- 在很多國家都適用，像是印度、巴西、加拿大等。（It is applicable

in many countries, including India, Brazil, Canada, and others.）

- 全世界有越來越多的建築物申請 LEED 的趨勢。（There is a growing trend worldwide for buildings to apply for LEED certification.）

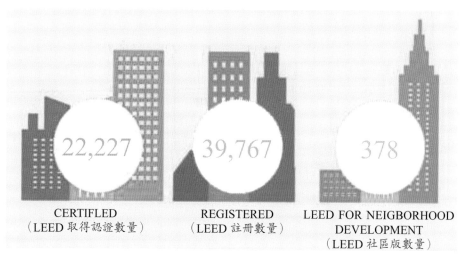

CERTIFLED
（LEED 取得認證數量）

REGISTERED
（LEED 註冊數量）

LEED FOR NEIGBORHOOD
DEVELOPMENT
（LEED 社區版數量）

來源：USGBC 統計資料

圖 1-8　LEED 案件數量

LEED 案件分成下列 2 大類別：（LEED projects fall into two main categories:）

- LEED 註冊專案：LEED 專案註冊是表示某個 LEED 專案已經透過線上申請形式申請了候選資格。（LEED Registered Projects: A LEED registered projcct indicates that the project has applied for candidate status through an online application process.）
- LEED 認證專案：LEED 認證表示專案已經成功被 GBCI 認可和得到 LEED 認證。（LEED Certified Projects: LEED certification indicates that the project has been successfully recognized and certified by the Green Business Certification Inc. (GBCI).）

　　而註冊專案類似於候選資格，認證專案類似於正式資格。（In this structure, registered projects are similar to candidate status, while certified projects are equivalent to formal certification.）

第二章　LEED 綠建築評估系統簡介（Introduction to the LEED Green Building Rating System）

學習目標（Learning Objectives）

- 什麼是 LEED（What is LEED）
- LEED 的原則（Principles of LEED）
- 各種 LEED 評分體系（Various LEED rating systems）
- LEED 各類評級系統是如何進行作業（How each LEED rating system operates）
- 美國綠色建築委員會（USGBC）、GBCI 與 LEED Online（U.S. Green Building Council (USGBC), GBCI, and LEED Online）
- LEED 認證的最低條件（MPR）（Minimum Program Requirements (MPR) for LEED certification）

2.1 USGBC 簡介（Introduction to USGBC）

　　美國綠建築協會（U.S. Green Building Council, USGBC）成立於 1993 年，是個非營利組織，旨在推動建築物能夠具有永續設計與建造。美國綠建築協會以推動領先能源與環境設計而著稱。（The U.S. Green Building Council (USGBC), founded in 1993, is a nonprofit organization aimed at promoting sustainable building design and construction. USGBC is most well-known for developing and promoting the LEED (Leadership in Energy and Environmental Design) certification system, which encourages energy-efficient and environmentally friendly building practices across the globe. The organization's mission is to transform the way buildings and communities are designed, built, and operated to improve the environmental and human health outcomes associated with built environments.）

LEED 領先能源與環境設計是美國綠建築協會在 2000 年設立的一項綠建築評分認證系統，用以評估建築績效是否能符合永續性，也提供 LEED 相關教育資源。（LEED (Leadership in Energy and Environmental Design) is a green building rating system established by the U.S. Green Building Council (USGBC) in 2000. It is designed to assess a building's sustainability performance and ensure that it meets specific environmental and energy efficiency standards. LEED not only evaluates the sustainability of buildings but also provides educational resources to help promote best practices in green building design, construction, and operation.）

這套標準逐步修正，目前適用版本爲 V4.1（2018）版本。適用建物類型包含；新建案、既有建築物、商業建築內部設計、學校、租屋與住家等。對於新建案（LEED NC），評分項目包括 7 大指標：選址與交通（Location and Transportation）、永續性基地（Sustainable Site）、用水效率（Water Efficiency）、能源和大氣（Energy and Atmosphere）、材料和資源（Materials and Resources）、室內環境品質（Indoor Environmental Quality）、創新和設計過程（Innovation and Design Process）、區域優先性（Regional Priority）。

（LEED provides evaluation criteria to ensure that buildings meet sustainability goals, and it also offers educational resources related to green building practices. The system has been continually updated, with LEED v4.1 (released in 2018) being the most current version. LEED applies to various building types, including new construction, existing buildings, commercial interiors, schools, rental properties, and residential homes. For new construction (LEED NC), projects are evaluated on seven key criteria:Location and Transportation、Sustainable Sites、Water Efficiency、Energy and Atmosphere、Materials and Resources、Indoor Environmental Quality、Innovation and Design Process、Regional Priority。）

圖 2-1　LEED 主要指標

　　評分系統中，總分為 110 分。申請 LEED 的建築物，如評分達 40-49，則該建築物為 LEED 認證級（Certified）；評分達 50-59，則該建築物達到 LEED 銀級認證（Silver）；如評分達 60-79，則該建築物達到 LEED 金級認證；如評分達 80 分以上，則該建築物達到 LEED 白金級認證（Platinum）。（In the LEED rating system, the total score is 110 points. Buildings applying for LEED certification are classified into different certification levels based on their score:LEED Certified: Buildings that score between 40-49 points; LEED Silver: Buildings that score between 50-59 points.LEED Gold; Buildings that score between 60-79 points; LEED Platinum: Buildings that score 80 points or more.）

目前台灣已取得 LEED 認證的建物爲辦公總部大樓，如：台北 101（LEED EBOM 白金級）、遠雄集團信義區 03 金融大樓（LEED NC 黃金等級）以及眾多高科技廠房。近兩年來，針對醫療院所類的 LEED HC（Healthcare），在台灣逐漸受到各醫療體系的重視，可能藉此作爲醫療環境高品質以及綠色的象徵。（Several buildings in Taiwan have obtained LEED certification, including Taipei 101 (LEED EBOM Platinum), Farglory Group's Xinyi District 03 Financial Building (LEED NC Gold), and many high-tech factories. In recent years, LEED Healthcare (HC) certification has gained attention in Taiwan's healthcare sector as a symbol of high-quality and sustainable healthcare environments.）

LEED 是 Leadership in Energy and Environmental Design（能源和環境設計領導）的首字母縮寫。它是國際公認的綠色建築評級系統，LEED 由美國綠建築協會（U.S. Green Building Council, USGBC）於 1994 年開始制定，1999 年正式公佈第一版本並接受評估申請。LEED 目前已成爲全美國各州公認之綠建築評估準則，各地方政府陸續將取得 LEED 認證標章列爲公共建設之必要條件，近年來更廣爲全世界其他先進國家所採用，或當作該國制定綠建築評估系統之範本。（LEED (Leadership in Energy and Environmental Design) was developed by the U.S. Green Building Council (USGBC) in 1994 and launched in 1999. It is now widely adopted globally as a standard for green building evaluation.）

　　LEED 旨在為業主和經營者提供簡明架構，確定並採取實際的、可衡量的綠色建築設計、施工、營運和維護的解決方案。（LEED provides a framework for building owners and operators to implement measurable green building design, construction, operations, and maintenance solutions.）

1. LEED 的特點是（Key Features of LEED:）

- LEED 提供了一個第三方公正的驗證，包括建築物或社區的設計與建造，以及以可持續營運的方式。（LEED provides third-party, impartial verification for the design, construction, and operation of buildings and communities in a sustainable way.）

- 自願認證制度。（It is a voluntary certification system.）

- 開發共同認可的制度與方法。（It develops a commonly recognized framework and methodology for green building practices.）

2. LEED 評級系統的發展基於三重底線（3Ps）（LEED rating system is based on the Triple Bottom Line (3Ps)）

Economic（經濟）= Profit

Enhanced Quality of Life

Environment（環境）= Planet　　Social（人文社會）= People

圖 2-2　LEED 評級系統的發展（3Ps）

- 社會責任（人 People）：為人提供更好的生活條件。（People (Social

Benefits) – Promoting occupant health, well-being, and community benefits.）

- 環境永續（地球 Planet）：對環境的影響較小。（Planet (Environ-
mental Benefits) – Reducing environmental impacts, resource consumption, and carbon emissions.）

- 經濟繁榮（利潤 Profit）：降低建築物的生命週期成本。（Profit
(Economic Benefits) – Lowering operational costs, increasing building
value, and fostering innovation through sustainable practices.）

2.2 各種 LEED 評級系統（Various LEED Rating Systems）

2.2.1 LEED 評分系統家族（LEED Rating System Families）

有五個 LEED 評分系統如下圖所述的家族類別。而這些評級系統是
適用於不同的專案類型（如下所述）。在 LEED V4 中，總共被歸類爲 21
個分類。（There are five LEED rating system families, as shown in the diagram
below. These rating systems apply to different project types (as described). In
LEED v4.1, they are categorized into a total of 21 subcategories.）

LEED 評級系統家族及其應用（LEED Rating System Families and Their Applications）

LEED for BD+C	LEED for ID+C	LEED for O+M	LEED for ND	LEED for HOMES
• 適用於新建大樓的建築，或是大範圍的更新。（Suitable for newly constructed buildings or large-scale renovations.）	• 適用的專案範圍僅限於內部裝修。（Limited to interior fit-out projects.）	• 正在進行更改的既有建築。 • 可能包括小幅的升級。（For existing buildings undergoing changes, possibly including minor upgrades.）	• 適用於土地開發專案。（Applicable to land development projects.） • 混合用途開發等。（Includes mixed-use developments.）	• 適用於所有住宅專案，除了高層住宅。（Applies to all residential projects, except high-rise residential buildings.）

　　這些評級系統包括針對各種專案類型的分類，確保每個系統都適用於特定的建築和開發專案，從住宅、商業建築到社區開發。（These rating systems include classifications for diverse project types, ensuring that each system is relevant to specific building and development projects, from residential to commercial and neighborhood development.）

圖 2-3　LEED 評估系統

主要裝修包括以下任一情況：（Major renovations include any of the following situations:）

- 建築外殼結構的改變（Changes to the building envelope: This includes structural alterations to the exterior, such as walls, roofs, or windows.）

- 冷暖通空調 HVAC 系統的變化（Modifications to the HVAC system: Significant updates or replacements of heating, ventilation, or air conditioning systems to improve energy efficiency or performance.）

- 主要室內裝修改變（Substantial interior renovations: This involves significant changes to interior finishes, layout, or functionality, affecting the core design of the building's interior.）

2.2.2 採用評分系統對不同的專案類型（Scoring systems apply differently based on project types）

LEED for Building Design and Construction
（建築設計與施工版）

- LEED BD+C: New Construction（新建）
- LEED BD+C: Core and Shell（核心及殼構造）
- LEED BD+C: Schools（學校）
- LEED BD+C: Retail（零售）
- LEED BD+C: Healthcare（醫療設施）
- LEED BD+C: Data Centers（資料中心）
- LEED BD+C: Hospitality（醫院）
- LEED BD+C: Warehouses and Distribution Center（倉庫）

LEED for Homes
（住宅）

- LEED Homes: Homes and Multifamily Lowrise（低密度住宅）
- LEED Homes: Multifamily Midrise（中密度住宅）

LEED for Interior Design and Construction
（宅內裝修版）

- LEED ID+C: Commercial Interiors（商業內裝）
- LEED ID+C: Retail（零售）
- LEED ID+C: Hospitality（醫院）

LEED for Building Operation and Maintenance
（營運和維護版）

- LEED O+M: Existing Buildings（既有建築物）
- LEED O+M: Data Centers（資料中心）
- LEED O+M: Warehouses and Distribution Centers（倉庫）
- LEED O+M: Hospitality（醫院）
- LEED O+M: Schools（學校）
- LEED O+M: Retail（零售）

LEED for Neighbourhood Development
（社區開發版）

- LEED ND: Plan（計畫）
- LEED ND: Build Project（專案建造）

NEW CONSTRUCTION & MAJOR RENOVATIONS　EXISTING BUILDINGS OPERATIONS & MAINTENANCE　COMMERCIAL INTERIORS　CORE AND SHELL DEVELOPMENT　RETAIL

SCHOOLS　HOMES　NEIGHBORHOOD DEVELOPMENT　HEALTHCARE　Selection Guidance

圖 2-4　LEED 評分系統之列表

1. LEED 建築設計和施工（LEED for Building Design and Construction）

　(1) LEED BD＋C：新建（New Construction）

　　商業用房、機構用房（博物館、教堂）和高層住宅樓，即居住建築高於九層樓適合居住的單元。（Commercial, institutional (e.g., museums, churches), and high-rise residential buildings (above nine stories).）

　(2) LEED BD＋C：結構體（Core & Shell）

　　整體建築中有超過承租人無法控制的設備或空間（商場、辦公大樓、倉庫等）。（Buildings where over 40% of the space is not under tenant control, such as malls, office buildings, or warehouses.）

　(3) LEED BD＋C：學校（Schools）

　　• K-12的學校教學樓將有申請 LEED BD＋C：學校這個類別的資格。
　　　（K-12 educational buildings.）

　　• 學校、大專教學大樓和幼稚園建築非教學樓可以依照專案團隊決定要申請 LEED BD＋C：學校或 LEED BD+C：NC。

　(4) LEED BD＋C：零售（Retail）

　　像銀行、餐館、服飾店、大型賣場等零售建築項目。（Projects like banks, restaurants, and stores.）

　(5) LEED BD＋C：醫療保健（Health Care）

　　• 住院和門診醫療設施（Inpatient and outpatient medical facilities.）

　　• 許可的長期保健設施

　　• 醫療機構、輔助生活設施與醫療教育和研究中心

　(6) LEED BD＋C：資料中心（Data Centers）

　　專門設計和裝備，以滿足高密度電腦設備的需求，如伺服器機架，用於數據的儲存和處理。（Buildings equipped for high-density computing equipment.）

(7) LEED BD＋C：旅館（Hospitality）

　　專用於提供過渡性或短期住宿或沒有餐廳的服務行業，包括飯店、汽車旅館、旅館或其他類似的產業。（Hotels, motels, and other short-term lodging.）

(8) LEED BD＋C：倉庫和配送中心（Warehouse and Distribution Centers）

　　用來囤貨，放置製造的產品、商品、原材料或個人財物。（Designed for storing products, raw materials, or personal belongings.）

2. LEED 住宅（LEED for Homes）

(1) LEED 住宅：住宅和低密度住宅（Homes and Multifamily Low-rise）

- 單戶住房（Single-family homes and low-rise residential buildings (up to 3 units).）
- 低密度多戶住宅型（最多 3 個可居住的單位）

(2) LEED 住宅：中密度住宅（Multifamily Midrise）

　　一層樓有 4 至 8 可居住單位以上的多戶型住宅建築。（Residential buildings with 4 to 8 units per floor.）

3. LEED 室內設計裝修（LEED for Interior Design and Construction）

(1) LEED ID＋C：零售（Retail）

　　如銀行、餐館、服飾店、大型零售等建築類別。（Banks, restaurants, stores.）

(2) LEED ID＋C：旅館（Hospitality）

　　室內空間，致力於提供過渡性或短期住宿或沒有餐廳之服務行業的酒店、汽車旅館、旅館或其他企業。（Hotels and motels.）

(3) LEED ID+ C：商業室內設計（Commercial Interiors）

適用於用戶空間，不包括零售和旅館專案。（Office spaces, excluding retail and hospitality.）

4. LEED 建築使用和維護（LEED for Operations and Maintenance）

(1) LEED O＋M：既有建築（Existing Buildings）

適用於可持續運行和維護建築物的既有建築，還應該包括系統升級與次要空間使用的改造。（Sustainable building operations, including system upgrades.）

(2) LEED O＋M：零售（Retail）

既有建築是使用要進行零售消費產品的場所，包括直接客戶服務領域（賣場區），以及為了服務準備或存儲區域（倉庫）。（Buildings used for retail purposes, including stores and storage areas.）

(3) LEED O＋M：學校（Schools）

K-12 以下學校的教學樓。也可用於大學校園和非教學樓。（Applies to K-12 educational buildings and can also be used for non-academic buildings on university campuses.）

(4) LEED O＋M：旅館（Hospitality）

服務行業提供過渡性或短期住宿或的現有飯店、汽車旅館、旅館或其他企業。（Covers existing hotels, motels, and similar enterprises offering short-term or transitional accommodations.）

(5) LEED O＋M：資料中心（Data Centers）

現有建築物特別設計和裝備，以滿足高密度電腦設備的需求，如伺服器機架，用於數據儲存和處理。（Applies to existing buildings designed to support high-density computing equipment like server racks for data storage and processing.）

(3) LEED O＋M：倉庫和配送中心（Warehouse and Distribution Centers）

用來存放貨物及製造的產品、商品、原材料或個人物品的既有建築物。（For existing buildings used to store goods, products, raw materials, or personal items.）

5. LEED 社區發展（LEED for Neighborhood Development）

(1) LEED ND：計畫（Plan）

在概念性規劃或正在建設的總體規劃階段的專案。（Projects in the conceptual or early construction planning stages.）

(2) LEED ND：專案建設（Build Project）

適用於已完成的開發專案。（Applicable to fully developed projects.）

2.2.3 LEED 認證人員的分級（LEED Certification Levels for Professionals:）

圖 2-5　認證人員分級示意

第一級：**LEED GA**——助理級，熟悉綠色策略、LEED 體系和認證流程。熟悉 LEED 家族體系關係和區別，什麼樣的建築選用什麼樣的認證體系；其次對各個得分點的策略也有把握，能夠清楚表達綠色建築的一些策略，並知道在專案中應用，且能支持一個專案達到 LEED 的設計標準，可以在專案團隊中發揮很大的作用；對 LEED 認證過程熟悉，特別是熟悉 LEED 專案的申請，所需資料、遞交、澄清、認證以及 LEED online 上的各種工具。（LEED Green Associate (GA): This is the entry-level certification for individuals who are familiar with green strategies, the LEED system, and the certification process. LEED GA professionals understand the various LEED systems and how to apply them to different building types. They can effectively support projects in achieving LEED design standards and are well-versed in the LEED online tools and application process.）

第二級：**LEED AP**（分 **BD + C**、**ID + C**、**HOME**、**ND**、**O&M** 五大類）——專業級，除了對 LEED GA 的內容完全掌握外，對 LEED 各個得分點的策略具有更深入、更具體的了解，能夠兼顧各方面的考慮，LEED AP 體現在技術細節、計算、規範理解、資源以及一定的專案管理協調能力。能夠獨立領導一個專案的諮詢和協調工作，當然也需要背後 LEED GA 和模擬工程師的支持。（LEED Accredited Professional (AP): There are five main categories of LEED AP, including BD+C, ID+C, Homes, ND, and O+M. A LEED AP not only has comprehensive knowledge of LEED Green Associate content but also understands the strategies for earning specific LEED credits. They are skilled in technical details, calculations, standards interpretation, resources, and project management. A LEED AP can independently lead project consultation and coordination, with the support of LEED GAs and simulation engineers.）

　　第三級：LEED Fellow——大師級，從事綠色建築行業十年以上，並具有 LEED AP 八年以上，對這個行業有一定的貢獻和了解，可以說是綠色建築行業的資深級顧問。申請 Fellow 有技術層面的要求，除了對內外部培訓技能的要求，還有職業領導能力等不同層次的要求。這個級別已經在 2011 年 9 月接受提名，第一批有 34 位綠色建築行業的人獲得 LEED Follow 提名，目前台灣僅一人獲得 Fellow 頭銜。（LEED Fellow: This is the expert level for professionals who have been in the green building industry for over ten years and have held LEED AP status for more than eight years. LEED Fellows are recognized for their significant contributions to the field and in-depth knowledge. To become a Fellow, there are technical and leadership requirements, including internal and external training skills. The first batch of LEED Fellows was nominated in September 2011, with 34 industry leaders recognized. Currently, only one person in Taiwan has achieved this title.）

　　對於想入行綠色建築又沒有綠色建築方面經驗的學員，建議先從考 LEED GA 開始，循序漸進。LEED GA 已經足夠讓你對 LEED 體系和流程有很大程度的了解，等有了經驗累積，在這個基礎上進階成 AP 不是什麼難事。（For those new to the green building field without prior experience, it is recommended to start with the LEED Green Associate certification and gradually progress to LEED AP once experience has been accumulated.）

2.3 如何選擇適當的評分系統（How to Choose the Appropriate Rating System）

　　一個專案若想申請 LEED 認證，需選擇最合適自身的評估體系。選擇評估體系的時候，可透過「40/60 原則」進行判斷：（When applying for LEED certification, a project must select the most suitable rating system. The choice can be determined using the "40/60 rule":）

　　如果某評估體系所適用的面積占總建築面積的 40% 以下，則不應採用這一體系。如果某評估體系所適用的面積占總建築面積的 60% 以上，則應採用這一體系。如果某評估體系所適用的面積占總建築面積的40%～60%，則專案團隊應作出獨立的評估，決定是否採用這一體系。（If the rating system applies to less than 40% of the total building area, that system should not be used.If the rating system applies to more than 60% of the total building area, that system should be used.If the rating system applies to 40%–60% of the building area, the project team should make an independent assessment to decide whether to use that system.）

　　截至 2023 年，全球已有超過 10 萬棟商業建築獲得 LEED 認證，並且有來自超過 180 個國家的專案正在申請 LEED 認證。僅在 2023 年，全球就有超過 6,000 個 LEED 商業專案獲得認證，這反映出國際上對於永續和環保建築的需求日益增加。（As of 2023, more than 100,000 commercial buildings around the world have achieved LEED certification, and there are projects in over 180 countries that are working toward LEED certification. In 2023 alone, over 6,000 LEED commercial projects were certified globally, reflecting increasing international demand for sustainable and environmentally conscious buildings.）

　　大多數專案可以使用一個特定的評價系統，但是在某些的情況下，專案是適用於一個以上的評級系統。以下策略將在這種情況下使用：（Most projects will be eligible for a specific rating system. However, in cases where a project could apply to more than one system, the following strategy should be applied:）

　　• 應用 60/40 規則，即專案中若 60% 以上適用於某個系統，則應該

選擇。（Use the 60/40 rule: If 60% or more of the project is applicable to one system, that system should be chosen.）

- 若是介於 40%~60% 之間，則由專案團隊選擇適合的評級系統。
（For projects where 40%–60% is applicable, the project team should decide which rating system is most appropriate.）

根據建築面積比例選擇

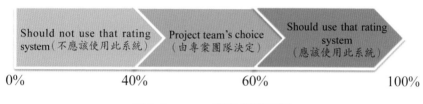

圖 2-6　LEED 評分選擇系統

2.4 了解 LEED 的評級系統（Understanding the LEED Rating System）

LEED 評分系統包括多個得分類別。在大多數的 LEED V4.1 評級系統是採用個別類別的得分加總計算出總分。（The LEED rating system is composed of multiple credit categories. In most LEED v4.1 rating systems, the total score is calculated by summing up the points from the individual credit categories.）

1. 而上述每一個得分分類別有三個要素組成：（Each credit category is comprised of three key elements:）

- 先決條件（或稱必要條件）（Prerequisites）
- 得分項目（得分點）（Credits (Points)）
- 分數（Score）

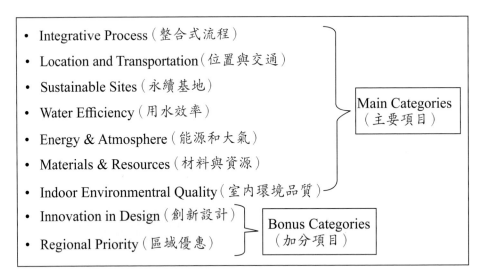

- Integrative Process（整合式流程）
- Location and Transportation（位置與交通）
- Sustainable Sites（永續基地）
- Water Efficiency（用水效率）
- Energy & Atmosphere（能源和大氣）
- Materials & Resources（材料與資源）
- Indoor Environmentral Quality（室內環境品質）
- Innovation in Design（創新設計）
- Regional Priority（區域優惠）

Main Categories（主要項目）

Bonus Categories（加分項目）

圖 2-7　LEED 評估系統主要項目與加分項目

　　由下圖 LEED 得分表中可以看出他是由先決條件（Prerequisites）、得分條件（Credit）與分數（Point）幾個部分所組成：（As illustrated in the LEED scorecard, the rating system is composed of three main parts: prerequisites, credits, and points.）

圖 2-8　LEED 之先決條件與得分條件

2. 先決條件（必要條件）是強制性的要求被認證的項目。只要申請 LEED 的專案，不論申請等級爲何，都一定需要被滿足的條件。（Prerequisites: These are mandatory requirements that a project must meet to be eligible for LEED certification. No points are awarded for fulfilling prerequisites, but they are necessary for the project's qualification.）

3. 得分條件是可選的項目。（Credits: These are optional sustainability strategies that contribute points to the total score. Different credits address various aspects of sustainable design, construction, and operations.）

4. 分數是當完成這個項目可以取得在 LEED 認證中的得分。（Points: Each credit has a point value based on its impact. The total points accumulated from all the credits determine the certification level (Certified, Silver, Gold, or Platinum).）

例如：參見上圖，在永續基地類別，施工汙染防治是一項先決條件。專案應符合這一要求，以獲得 LEED 認證。而其他像是基地評估、開放空間等其他要求是可選擇的。通過滿足得分項目能夠獲得一定的分數。（For example, as shown in the diagram above, in the Sustainable Sites category, Construction Pollution Prevention is a prerequisite. The project must meet this requirement to be eligible for LEED certification. Other criteria, such as Site Assessment and Open Space, are optional. By fulfilling these optional credits, the project can earn a certain number of points.）

- 根據得分條件的數量，該專案可能被評爲 LEED 認證級、LEED 銀級認證、LEED 金級認證或 LEED 白金級認證。（Based on the number of points earned, the project may be awarded LEED Certified, LEED Silver, LEED Gold, or LEED Platinum certification.）

40-49 分　　　50-59 分　　　60-79 分　　　80 分以上
認證級　　　　銀級　　　　　金級　　　　　白金級

圖 2-9　LEED 證書之等級

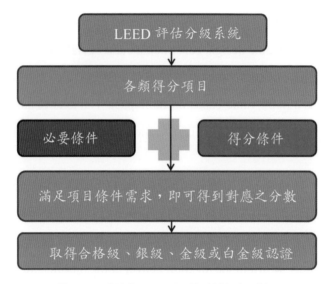

圖 2-10　綜述 LEED 評分系統如何進行

2.5 LEED 的影響類別及比重（LEED Impact Categories and Weightings）

參照下圖，所有的分數都具有不同的比重。分數分配是由於 LEED 的目標分配所造成。（As shown in the table, all points are weighted differently based on LEED's impact goals. The point distribution reflects the emphasis placed on different sustainability objectives.）

　　LEED 認證的建築，預計有以下積極的影響：（LEED-certified buildings are expected to have the following positive impacts:）

- 減少對全球氣候變化的危害（Reduction in harm to global climate change）
- 提高人類健康（Improvement of human health）
- 保護和恢復水資源（Protection and restoration of water resources）
- 保護和加強生物多樣性和生態系統（Protection and enhancement of biodiversity and ecosystems）
- 促進可持續發展和再生物質循環（Promotion of sustainable development and the regeneration of natural materials）
- 建立綠色經濟（Support for a green economy）
- 提升生活品質的社區（Creation of communities that enhance quality of life）

　　這些 LEED 目標被稱為影響類別。每一個影響類別根據他們的重要性而給予不同的權重。例如與氣候變化相比，綠色經濟就更為重要了。

（LEED Impact Categories and WeightingsAs shown in the table, all points are weighted differently based on LEED's impact goals. The point distribution reflects the emphasis placed on different sustainability objectives. LEED-certified buildings are expected to have the following positive impacts:Reduction in harm to global climate changeImprovement of human healthProtection and restoration of water resourcesProtection and cnhancement of biodiversity and ecosystemsPromotion of sustainable development and the regeneration of natural materialsSupport for a green economyCreation of communities that enhance quality of lifeThese LEED goals are known as impact categories. Each impact category is weighted according to its importance. For example, addressing climate change may be given more weight than promoting a green economy.）

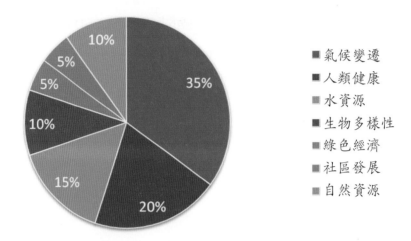

■ 氣候變遷
■ 人類健康
■ 水資源
■ 生物多樣性
■ 綠色經濟
■ 社區發展
■ 自然資源

圖 2-11　LEED 之不同影響的權重類別

2.6 LEED 參考手冊（LEED Reference Manuals）

　　LEED 參考手冊是由美國綠色建築委員會（USGBC）出版，包含詳細特定 LEED 評分系統的所有技術問題及細節。LEED 參考手冊詳細介紹了目的、要求、策略、計算方式等。為每個分數／必要條件提供了明確的規範。（LEED Reference Manuals are published by the U.S. Green Building Council (USGBC) and contain all the technical details and specific requirements of the LEED rating systems. The manuals provide comprehensive information on objectives, requirements, strategies, and calculation methods for each category. They also include clear guidelines for earning points and meeting prerequisites. These reference guides serve as an essential resource for project teams to ensure they meet LEED certification standards.）

圖 2-12　LEED 參考手冊

2.7 美國綠建築協會（USGBC）、綠建築認證委員會（GBCI）及 LEED 線上（LEED Online）（U.S. Green Building Council (USGBC), Green Business Certification Inc. (GBCI), and LEED Online）

　　USGBC，美國綠色建築協會，由其開發的 LEED 為各種各樣複雜的建築體系提供了一個可衡量的標準，並得到了廣泛成功的應用。其特點是以委員制立基，以會員制推動，以投票制聚集。現在已超過 20,000 個會員組織，15 萬以上 LEED 資格人員。美國綠色建築委員會（USGBC）側重於制定和完善 LEED 的標準體系、宣傳教育及綠色工作等內容。(US-GBC (U.S. Green Building Council) is the organization that developed LEED (Leadership in Energy and Environmental Design), a measurable standard for assessing the sustainability of various types of complex building systems. USGBC operates on a committee-based system, driven by its members, with decisions made through a voting process. USGBC has grown to over 20,000 member organizations and more than 150,000 LEED-accredited professionals. Its primary focus is on developing and refining the LEED standard, promoting green education, and advocating for sustainable practices.）

GBCI，綠色建築認證委員會，成立於 2008 年，主要協助 USGBC 完成 LEED 建築認證和 LEED 資格認證，是建築和人才兩類認證基準的審核機構。（GBCI (Green Business Certification Inc.) was established in 2008 to assist USGBC in administering LEED certification for buildings and accreditation for professionals. GBCI functions as the certifying body for both building projects and individual LEED professionals, ensuring that the standards are met.）

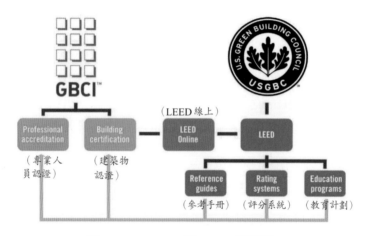

圖 2-13　USGBC 與 GBCI 關係圖

美國綠色建築協會、GBCI 和 LEED 的主要角色說明如下：（The roles of the U.S. Green Building Council (USGBC), Green Business Certification Inc. (GBCI), and LEED Online are outlined as follows:）

- 美國綠色建築協會負責發展評級系統，參考手冊和教育計畫。（USGBC is responsible for developing the rating systems, reference manuals, and educational programs.）
- GBCI 管理建築認證和專業人員認證。（GBCI manages both building certification and professional accreditation.）

- LEED online 是線上申請工具，包含整個 LEED 認證申請流程。

  （LEED Online is the online application tool that facilitates the entire LEED certification process. It stores all project details (scorecards, drawings, supporting documentation, etc.) in an online system. Through LEED Online, project teams can access the CIR (Credit Interpretation Rulings) database, scoring systems, errata, and other resources.）

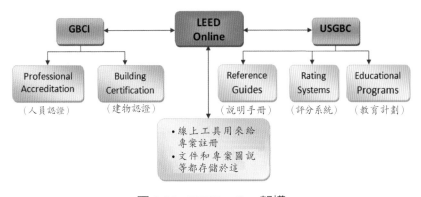

圖 2-14　LEED online 架構

- ■ 它包含所有專案的詳細訊息（得分卡、圖說、證明文件等）的 LEED 文件都儲存在一個線上儲存系統。（It contains all project details, such as scorecards, drawings, supporting documentation, etc., stored in an online system.）

- ■ 專案團隊通過 LEED online 可以進入 CIR 資料庫、評分制度、勘誤表等等。（Through LEED Online, project teams can access the CIR (Credit Interpretation Rulings) database, scoring systems, errata, and other resources.）

無論 USGBC 還是 GBCI，都使用 LEED online 聯繫。這意味著，每個用戶 ID 註冊 USGBC/ GBCI/ LEED online 都必須是獨一無二的，所以不

能和別的註冊使用者使用相同的用戶名和密碼，不論是 www.usgbc.org 或 www.gbci.org 或 www.leedonline.com。（Both USGBC and GBCI use LEED Online for communication. This means that every user ID registered with USGBC/GBCI/LEED Online must be unique. Users cannot share the same username and password for www.usgbc.org, www.gbci.org, or www.leedonline.com.）

其他值得注意的以下幾點：（Here are some important points to note:）

- LEED Green Associate 和 LEED AP 並不是證書（Certification）。他們被稱爲 LEED 認證（Credentials）。（LEED Green Associate and LEED AP are not certifications; they are referred to as credentials.）

- 個人是不能成爲美國綠建築協會的會員的。只有公司才可以申請爲會員。成爲美國綠建築委員會會員公司的全職員工可以利用會員資格的好處，包括會員公司可以在網站、名片上使用美國綠色建築委員會成員標誌和電子郵件簽名。但是不應該在任何產品上使用此標誌或包裝。（Individuals cannot become members of the U.S. Green Building Council (USGBC). Only companies can apply for membership. Full-time employees of USGBC member companies can benefit from the membership, such as using the USGBC member logo on websites, business cards, and email signatures. However, the logo should not be used on any products or packaging.）

- LEED 不認可或證明任何產品或材料。（LEED does not endorse or certify any products or materials.）

美國綠色建築委員會章標誌只能按照分會作爲官方業務的一部分（如分會出版物、網站等等）。它並不允許由分會成員或個人使用。（The USGBC chapter logo can only be used as part of official chapter business (e.g., chapter publications, websites, etc.). It is not permitted to be used by chapter members or individuals.）

　　同樣，美國綠色建築協會會員標誌僅可用於美國綠色建築協會的會員成員公司。其他需要使用 USGBC 標誌和 LEED 標識是由美國綠色建築協會市場部（聯繫 marketing@usgbc.org）根據實際情況逐案通過書面申請及審查。（Similarly, the USGBC member logo can only be used by member companies of the U.S. Green Building Council. Any other usage of the USGBC logo or LEED insignia must be approved on a case-by-case basis by the USGBC marketing department. For such requests, written applications must be submitted and reviewed by contacting marketing@usgbc.org.）

圖 2-15　USGBC 標章範例

　　關於美國綠色建築委員會：（About the U.S. Green Building Council (USGBC):）

- 美國綠色建築委員會是一個非營利的組織，透過低成本和節能節水型綠色建築物，希望能創造一個更好和永續性的未來。（USGBC is a nonprofit organization that aims to create a better and more sustainable future through low-cost, energy-saving, and water-efficient green buildings.）

- 任務：改造建築及社區的設計方式、建造和營運模式，實現對環境和社會負責且健康和繁榮的環境，以提高生活品質。（Mission: To transform the way buildings and communities are designed, con-

structed, and operated, promoting environmentally and socially responsible spaces that foster health and prosperity, ultimately improving the
quality of life.）

- 願景：建築和社區將重新再生且維持健康，保有所有生命生生不
息的一代。（Vision: Buildings and communities will be regenerative
and maintain health, supporting all life for generations to come.）

2.8 LEED 最低專案要求（Minimum Program Requirements）

　　每一個申請 LEED 的專案都必須符合最低專案要求（MPR），這是非
常基本的條件，因此應該滿足此基本要件才有資格註冊爲 LEED 認證。
（Each project applying for LEED certification must meet the Minimum Program Requirements (MPRs), which are the basic conditions that must be satisfied
to qualify for LEED certification registration.）

1. 一般要求（General Requirements:）

- 必須在現有的永久土地位置上。（The project must be located on existing, permanent land.）
- 必須使用合理和一致的基地邊界。（It must use a reasonable and consistent project boundary.）
- 必須遵守專案要求的面積大小。例如 LEED BD + C 和 LEED O +
M 需要 1000 平方英尺以上面積，而 LEED ID + C 需要 250 平方
英尺以上的面積。（The project must comply with minimum floor area
requirements, such as:LEED BD+C and LEED O+M require at least
1,000 square feet.LEED ID+C requires at least 250 square feet.）

2. LEED 社區發展的最低要求（Minimum Requirements for LEED Neighborhood Development:）

　　LEED 專案中應包含至少兩個可居住的住宅，並不得大於 1500 英畝。
（The project must include at least two habitable residential units and should not
exceed 1,500 acres.）

3. LEED 住宅的最低要求（Minimum Requirements for LEED Homes:）

- LEED 專案必須由適用的法規所定義為「住宅單位」。（The project
must be defined as a "dwelling unit" by applicable codes.）

　　該 LEED 專案邊界提交是屬於 LEED 認證的部分。對於多個不同用
途的 LEED 專案在同一個基地上來說，專案的邊界是可以由專案小組決
定的。（The submitted LEED project boundary is the part that applies for LEED
certification. For multiple projects with different uses on the same site, the project boundary can be determined by the project team.）

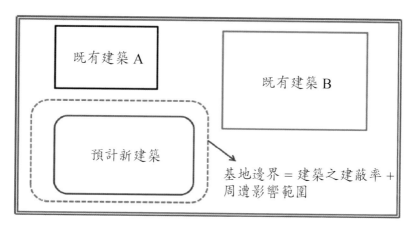

圖 2-16　基地界限／LEED 邊界

- MPR 的目的：（The purpose of the MPR (Minimum Program Requirements)）

■ 提供客戶明確的指導方針（Provide clear guidelines for clients.）

■ 以減少認證過程中可能出現的挑戰及問題（Reduce potential challenges and issues during the certification process.）

■ 爲了保護 LEED 程序的完整性（Protect the integrity of the LEED certification process.）

如果該專案不符合 MPR 的條件，GBCI 可隨時撤銷認證。（If the project does not meet the MPR conditions, GBCI reserves the right to revoke the certification at any time.）

2.9 LEED 專案申請認證的要求（LEED Project Certification Requirements）

- 該專案應遵守各評級系統的最低專案要求。（The project must meet the Minimum Program Requirements (MPRs) of the selected rating system.）

- 該專案應遵守各評級系統的所有必要條件。（The project must comply with all prerequisites of the applicable rating system.）

- 該專案應獲得認證等級所需的點最低分數。（The project must achieve the minimum number of points required for the desired certification level.）

目前這些評分指標被分爲整合過程和八大方面的分數，細分如下。從下表可以看出每個章節所涵蓋的得分比重略有不同，也看出基地、能源爲占比最重的兩個項目。（Currently, the LEED scoring criteria are divided into integrative process and eight main categories, as shown below. The table illustrates the different weightings for each category, with site sustainability and energy being the two categories with the highest point allocation.）

整合過程（Integrative Process）　　　　　1 分（1%）

選址與交通（Loction & Transprtation）　　16 分（15%）

永續基地（Sustainable Stres）　　　　　10 分（9%）

用水效率（Water Efficiency）　　　　　11 分（10%）

能源與天氣（Energy and Atmisphere）　　33 分（30%）

材料與資源（Materiais and Resiurces）　　13 分（12%）

室內環境品質（Indoor Environmental Quality）　16 分（15%）

創新與設計（Innovation and Design Process）　6 分（6%）

地域優先（Regional Priority）　　　　　4 分（4%）

總分 110 分（100%）

- LEED 先決條件列表如下：（Here is the list of LEED prerequisites that projects must meet for certification:）

必要項目名稱（英文）	必要項目名稱（中文）
Prerequisite: Integrative Project Planning and Design (Healthcare)	先決條件：綜合項目規劃與設計（醫療）
SS Prerequisite: Construction Activity Pollution Prevention	SS 先決條件：施工活動汙染防治
SS Prerequisite: Environmental Site Assessment	SS 先決條件：環境現場評估
WE Prerequisite: Outdoor Water Use Reduction	WE 先決條件：戶外減少用水量

必要項目名稱（英文）	必要項目名稱（中文）
WE Prerequisite: Indoor Water Use Reduction	WE 先決條件：室內減少用水量
WE Prerequisite: Building-Level Water Metering	WE 先決條件：建築分水錶
EA Prerequisite: Fundamental Commissioning and Verification	EA 先決條件：基本調試和驗證
EA Prerequisite: Minimum Energy Performance	EA 先決條件：最低能源性能
EA Prerequisite: Building-Level Energy Metering	EA 先決條件：建築分電錶
EA Prerequisite: Fundamental Refrigerant Management	EA 先決條件：基本冷媒管理
MR Prerequisite: Storage and Collection of Recyclables	MR 先決條件：貯藏和回收的可回收物
MR Prerequisite: Construction and Demolition Waste Management Planning	MR 先決條件：廢棄物管理計畫
MR Prerequisite: PBT Source Reduction-Mercury (Healthcare)	MR 先決條件：PBT 源頭減量，汞（醫療保健）
EQ Prerequisite: Minimum Indoor Air Quality Performance	EQ 先決條件：最低室內空氣品質性能
EQ Prerequisite: Environmental Tobacco Smoke Control	EQ 先決條件：環境煙害控制
EQ Prerequisite: Minimum Acoustic Performance	EQ 先決條件：最低聲學性能

　　在接下來的章節會學習到，在下面的每個類別介紹每個先決條件和得分條件：（In the upcoming sections, you will learn about each prerequisite and credit in the following categories:）

- 基地位置和交通（Location and Transportation）

- 永續基地（Sustainable Sites）
- 用水效率（Water Efficiency）
- 能源與大氣（Energy & Atmosphere）
- 材料與資源（Materials & Resources）
- 室內環境品質（Indoor Environmental Quality）
- 創新設計（Innovation in Design）
- 區域優先（Regional Priority）

2.10 LEED4.0 版本及 LEED4.1 版本的差異與更新（LEED v4.0 and LEED v4.1 Key Differences）

LEED v4.1 於 2018 年推出，作為對 LEED v4 的漸進式更新。其測試版於 2018 年 3 月發布，並在 2019 年第一季度進行了公開意見徵詢。LEED v4.1 專注於提高認證過程的靈活性和根據實際表現的評估，並更新了能源效率、水資源和室內空氣品質等要求。同時，該版本整合了通過 Arc 平台進行的性能追蹤，以便更有效地評估建築的永續性。（LEED v4.1 was introduced in 2018 as an incremental update to the previous LEED v4 version. The beta version was launched in March 2018, with the final version expected to go through official public commenting by the first quarter of 2019. LEED v4.1 focuses on making the certification process more flexible and performance-driven, with updated performance thresholds and reference standards to reflect industry advancements. It streamlines prerequisites for energy, water, and indoor air quality while integrating performance tracking through the Arc platform.）

LEED v4.0 和 LEED v4.1 的主要差異在於對可操作性、靈活性以及實際表現的強調，以下是兩者的幾個關鍵不同點：（LEED v4.0 and LEED v4.1 have some significant differences, primarily focused on improving flex-ibility, addressing newer industry standards, and making the certification process more achievable and inclusive. Below are some of the key differences:）

1. 基於表現的評估：（Performance-Based Approach:）

　　LEED v4.1 更強調基於表現的標準，特別是在能源和用水方面。專案需要展示實際的數據表現，而不僅僅是遵守預設的標準。這使得綠建築的永續性通過更具體的數據來評估。（LEED v4.1 places greater emphasis on performance-based standards, particularly in areas like energy and water. Proj-ects must now demonstrate performance outcomes rather than just comply with prescriptive requirements. This shift allows for more accurate, data-driven mea-surements of a building's sustainability.）

2. 能源效率：（Energy Efficiency:）

　　在 LEED v4.0 中，能源效率主要依賴於預測模型，而 LEED v4.1 更加側重於實際的能源表現，即專案必須在建造完成後展露眞實世界的能源使用數據。（In LEED v4.0, energy efficiency was based largely on modeling predicted energy usage, whereas LEED v4.1 incorporates actual energy perfor-mance. Projects need to demonstrate how well they perform after construction using real-world data.）

3. 材料與資源：（Materials and Resources:）

　　LEED v4.1 對於材料信用結構進行了簡化，提供了更多記錄材料的選擇，並且要求更靈活的材料透明度報告和成分揭示方式。（LEED v4.1 simplifies the credit structure around materials, offering new options for docu-menting materials and making the requirements more flexible. For example, it provides more pathways for material transparency and ingredient reporting.）

4. 住宅及既有建築：（Residential and Existing Buildings:）

LEED v4.1 對既有建築和住宅專案進行了更新，更加注重營運階段的表現，使這類建築更容易獲得認證，與 v4.0 相比，設計階段的要求不再是唯一的考量。（LEED v4.1 introduces updates for existing buildings and residential projects, making it easier for existing buildings to achieve certification by focusing on operational performance instead of just design.）

5. 簡化的文件要求：（Simplified Documentation:）

LEED v4.1 簡化了某些文件和要求，特別是在材料得分項目和基於表現的得分項目方面，這使得專案團隊更容易管理整個認證過程。（LEED v4.1 reduces the complexity of documentation and requirements, especially in terms of materials credits and performance-based credits, making it easier for project teams to manage the certification process.）

6. 區域適應性：（gional Adaptation:）

LEED v4.1 提供了更多的區域適應性選項，使其更能適用於不同地區的需求，而 v4.0 的要求相對更為統一。（ED v4.1 offers greater regional adaptability, allowing it to be more applicable across different geographic areas, unlike v4.0, which was more uniform in its requirements.）

這些更新使 LEED v4.1 更加使用者友善，並更能應對當今建築環境中管理永續專案的實際挑戰。這些變化反映了提高 LEED 認證流程的可用性和可及性的需求，從而鼓勵全球更廣泛的採用。（ese updates make LEED v4.1 more user-friendly and responsive to the practicalities of managing sustainable projects in today's construction landscape. The changes reflect a desire to improve the usability and accessibility of the LEED certification process, encouraging wider adoption globally）

第三章 整合流程、規劃與評估（Integrative Process, Planning, and Assessments, IP）

3.1 LEED V5 版本簡介（Introduction to LEED V5）

　　LEED v5 是全球公認的綠色建築實踐綜合框架的最新版本。為了響應市場對更高責任感的需求，LEED v5 將專注於推動脫碳、生態系統保護與恢復、公平性、健康和韌性等關鍵議題，從而推動全球範圍內的實際影響和積極變革。（LEED v5 is the next version of the globally recognized comprehensive framework for green building practices.Embracing market demands for greater accountability, v5 will champion solutions to align the built environment with critical imperatives including decarbonization, ecosystem conservation and restoration, equity, health, and resilience. LEED v5 will drive real-world impact and positive change worldwide.）

　　LEED v5 共有三大核心目標，得分比重如下：（LEED v5 has three core goals, with the following score distribution:）

- 氣候行動（Climate Action (50%)）
- 生活品質（Quality of Life (25%)）
- 生態保育與恢復（Ecological Conservation and Restoration (25%)）

三大核心目標 × 五項原則 （Three Core Goals × Five Principles）

LEED v5 的三大核心目標乍看之下有許多交錯與重疊之處，然而這便是 LEED v5 突破以往的新布局──垂直整合所有得分點。（At first glance, LEED v5's three core goals may appear to overlap, but this is part of its innovative layout—a vertical integration of all credits.）

USGBC 展示了下方表格，其中三大核心目標列於左側，欄位上方則為五項原則，並將得分點舉例填入表格中，以說明 LEED v5 所有得分點的共同目標。（USGBC presented the table at 2023, where the three core goals are listed on the left and the five principles across the top. Example credits were filled in to illustrate how all LEED v5 credits align with these common goals.）

	DECARBONIZATION	RESILIENCE	HEALTH AND WELL-BEING	EQUITY	ECOSYSTEMS
CLIMATE ACTION	Decarbonize swiftly	Total carbon and peak demand reductions slow the rate of climate change	• Indoor air improvements • Air and water quality improvements from total carbon reductions	• Energy burden • Air and water quality improvements from total carbon reductions	• Air and water quality improvements from carbon reductions • Ecosystem impacts of nuclear/hydro power • Other impact categories within an LCA
QUALITY OF LIFE	Prioritize passive vs. mechanical efficiency for passive survivability	Adapt for resilience	Reduce stress and impact of climate change events on people	Reduce and delay the impact of climate change on frontline communities	Value and prioritize ecosystems services that bolster resilience
	Improved indoor air quality from no internal combustion	Design for mental health resilience (comfort, wayfinding, acoustics)	Invest in human health and well-being	Consider health and well-being throughout the supply chain, on construction sites, transportation corridors and downstream	Value and prioritize ecosystem services that bolster human health and well-being
	Account for who causes climate change compared to who experiences it first and worst	Bolster resilience for the most vulnerable (e.g.: underhoused, incarcerated, frontline communities)	Community support and engagement	Create equitable outcomes	Equitable access to nature
ECOLOGY	Circularity/lifecycle analysis and reuse of materials	Reduce flooding and extreme heat/wildfires	Improved air and water quality, affordability	Mitigate urban heat island impacts	Restore ecosystems

（來源：USGBC）

而五項原則分別是：（The five principles are:）

- 脫碳（Decarbonization）
- 韌性（Resilience）

- 健康與福祉（Health and Well-Being）
- 公平（Equity）
- 生態系統（Ecosystems）

USGBC 在制定 LEED v5 版本目標時，希望爲市場提供一套全面的工具，以實現永續的脫碳經營目標。因此，LEED v5 以「朝向低碳、並著重公平性及生物多樣性」爲方向，透過三大原則──「生態保護與恢復」（Ecological Conservation & Restoration）、「生活品質」（Quality of Life）、和「氣候行動」（Climate Action）──發展出五大核心價值：（In setting the goals for LEED v5, USGBC aims to provide the market with a comprehensive toolset to achieve sustainable decarbonization targets. Therefore, LEED v5 is guided by the principles of "Low Carbon, with a Focus on Equity and Biodiversity," and develops five key categories based on three core principles: "Ecological Conservation & Restoration," "Quality of Life," and "Climate Action."）

- **脫碳（Decarbonization）**：迅速實現建築業脫碳，以反映氣候危機的迫切性。（Rapidly achieve decarbonization in the building sector to reflect the urgency of the climate crisis.）
- **彈性（Resilience）**：激發並認識適應性和具備彈性的建築環境。（Inspire and recognize adaptable and resilient built environments.）
- **健康和福祉（Health and Well-being）**：投資於人類健康和福祉。（Invest in human health and well-being.）
- **公平性（Equity）**：創造一個多元化、公平和包容性蓬勃發展的環境。（Create a diverse, equitable, and inclusive thriving environment.）
- **生態系（Ecosystems）**：透過再生性發展實踐支持繁榮的生態系統。（Support thriving ecosystems through regenerative development practices.）

建築設計和建造的時間通常爲 3 至 5 年，而建築運營壽命則可持續 20 至 30 年，因此「運營與維護」（O+M）尤爲重要。爲此，USGBC 計畫將 LEED v5 標準帶到第 28 屆氣候變遷會議（COP28），以推動建築成爲實現脫碳的重要途徑之一，並在考量經濟因素的同時，減少環境影響並提升生活品質。（Building design and construction usually take 3 to 5 years, but the operational life of a building can last 20 to 30 years, making Operations and Maintenance (O+M) particularly critical. Thus, USGBC plans to bring LEED v5 standards to the 28th Climate Change Conference, advocating for buildings as a key pathway to decarbonization, while considering economic factors to reduce environmental impacts and improve quality of life.）

LEED v5 公眾意見徵詢（LEED v5 public comment）

第二次 LEED 公眾意見徵詢：2024 年 9 月 27 日至 10 月 28 日

第一次 LEED 公眾意見徵詢：2024 年 4 月 3 日至 5 月 24 日

在公眾意見反饋期間，所有利益相關者都可以查看變更並提供意見，無任何限制。他們將根據收到的意見進一步完善草案，最終形成最新版本的 LEED。（During the public comment feedback periods, we seek all stakeholders' input—there are no restrictions on who may view the changes and provide feedback. This input will be incorporated as we refine the drafts into what will be the newest version of LEED.）

評級系統草案與資源（Rating system drafts）

• LEED BD+C：新建建築（LEED BD+C: New Construction）

• LEED BD+C：核心與外殼（LEED BD+C: Core and Shell）

• LEED ID+C：商業內部裝修（LEED ID+C: Commercial interiors）

• LEED O+M：既有建築（LEED O+M: Existing Buildings

時程與影響（Timeline and impact）

LEED v5 計畫於 2025 年初開放註冊，作為最新的評估系統。儘管關鍵日期可能會有所調整，但其目標保持不變：推出一個具前瞻性、可實現但更具挑戰的 LEED v5 計畫，最重要的是，該計畫將帶來深遠的影響。（Early 2025 is the target for opening LEED v5 for registration, as the newest balloted rating system. While milestone dates may shift over time, the goal remains the same: we are committed to releasing a LEED v5 program that is relevant and forward-thinking, that is attainable yet raises the bar. Above all, it will be impactful!）

LEED v5 的開發圍繞三大核心影響領域進行：脫碳、生活品質、生態保護與恢復。LEED v5 中的每個得分項目與必要條件都與脫碳、生活品質和／或生態保護相關，這些連結在評級系統中有所標註，方便專案團隊塑造其永續發展的故事並有效傳達。（LEED v5 has been developed around three central areas of impact: Decarbonization, Quality of life, Ecological conservation and restoration. Every credit and prerequisite in LEED v5 has a connection to decarbonization, quality of life and/or ecological conservation, and this is annotated throughout the rating system so that project teams can easily shape their sustainability stories and communicate them..）

LEED v5 與 LEED v4.1 的差異：（Differences between LEED v5 and LEED v4.1）

LEED v5 在整合性和垂直整合方面顯著提升，針對所有得分項目的目標進行整合，旨在創造更具永續性的建築。（LEED v5 offers significant improvements in integration and vertical alignment, aligning the objectives of all scoring items to create more sustainable buildings.）

相比之下，LEED v4.1 專注於提升能源效率和減少營運碳排放，而 LEED v5 則進一步解決了建築物所有主要的碳排放來源，並提供了具體步驟來應對這些挑戰。（In contrast, LEED v4.1 focused on energy efficiency and reducing operational carbon emissions, while LEED v5 addresses all major carbon emission sources in buildings and provides concrete steps to tackle these challenges.）

此外，雖然 LEED v4.1 只在一定程度上提及隱含碳，但 LEED v5 則要求專案深入研究和量化隱含碳排放。（Moreover, while LEED v4.1 mentioned embodied carbon to some extent, LEED v5 requires projects to deeply analyze and quantify embodied carbon emissions.）

LEED v5 也更強調脫碳、健康與福祉，以及生態系統的保護，並設立了專門的小組來處理建築韌性和社會公平等長期關注的議題。（LEED v5 places greater emphasis on decarbonization, health and well-being, and ecosystem protection, with dedicated task forces addressing long-standing concerns about building resilience and social equity.）

在公平性方面，LEED v5 採用一個以公平為核心的框架，並新增了相關得分項目，以推動創造包容性建築。（In terms of equity, LEED v5 adopts an equity-centered framework and introduces new credits to promote the creation of inclusive buildings.）另有新增的「健康與幸福感」得分項目，促進居住者的福祉。（The "Health and Well-being" credits have also been added to enhance occupant well-being.）

LEED v5 還賦予專案團隊更大的靈活性來累積得分項目，並更加重視創新技術，鼓勵專案反映最新的永續設計理念。（LEED v5 gives project teams greater flexibility in earning credits and places more emphasis on innovative technologies, encouraging projects to reflect the latest in sustainable design.）

其得分項目系統更加全面、相互關聯，旨在共同促進建築的永續發展。（Its scoring system is more comprehensive and interconnected, aiming to collectively drive sustainable building development.）

3.2 氣候韌性評估（Climate Resilience Assessment）

目標（Intent）

促進對已觀察到、預測和未來自然災害的全面評估，以提升對氣候韌性的認識，增加風險透明度，減少脆弱性，並確保長期的安全性和永續性。（To promote comprehensive assessment of observed, projected, and future natural hazards for climate resilience, aiming to enhance awareness of hazards, increase transparency of risks, reduce vulnerabilities, and ensure long-term safety and sustainability.）

在評估中，辨識可能影響專案地點和建築功能的已觀察到、預測和未來的自然災害。需考慮的基地特定自然災害包括但不限於乾旱、極端高溫、極端低溫、洪水、颶風和強風、冰雹、山崩、海平面上升和風暴潮、龍捲風、海嘯、野火及煙霧、冬季風暴及其他相關災害（具體說明）。（As part of the assessment, identify observed, projected, and future natural hazards that could potentially affect the project site and building function. Address site-specific natural hazards, including, but are not limited to, drought, extreme heat, extreme cold, flooding, hurricane and high winds, hail, landslide, sea level rise and storm surge, tornado, tsunami, wildfire and smoke, winter storm, and other relevant hazards (specify).）

Transformative actions and system transitions

(a) Societal choices that generate fragmented climate action or inaction and unsustainable development perpetuate business as usual development

(b) Societal choices that support CRD involve transformative actions that drive five systems transitions

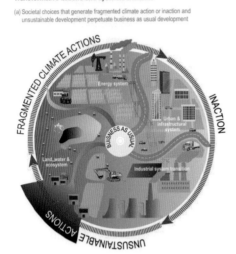

圖 3-1　氣候韌性發展路徑的特徵是變革性行動和系統轉型。(Transformative actions and system transitions characterize Climate Resilient Development Pathways.) 圖片來源：IPCC。

　　至少能辨識兩個優先災害，並提出設計策略以應對這些災害。對於每個優先災害，專案團隊必須評估並具體說明以下內容：(Identify two priority hazards, at minimum, to address through proposed design strategies. For each priority hazard, the project team must assess and specify the following:)

- 使用的 IPCC 排放情景，並說明共享社會經濟途徑 (IPCC emissions scenario used, specifying the Shared Socioeconomic Pathways)
- LEED 專案的預計使用壽命 (例如：至 2050 年或 100 年) (Projected service life of the LEED project (e.g., FY2050 or 100 years))
- 災害等級 (Hazard level)
- 災害風險評級 (Hazard risk rating)
- 暴露度、敏感性、適應能力、脆弱性和總體風險等級 (Exposure, sensitivity, adaptive capacity, vulnerability, and overall risk levels)

- 對專案地點和建築功能的潛在影響（Potential impact on the project site and building function）
- 對施工期間專案地點的潛在影響（Potential impact on the project site during construction）

　　在可能的情況下，使用評估資訊來指導專案的規劃、設計、運營和維護，並描述如何考慮專案特定的策略。（Where possible, use the information from the assessment to inform the planning, design, operations, and maintenance of the project and describe how project-specific strategies were considered.）

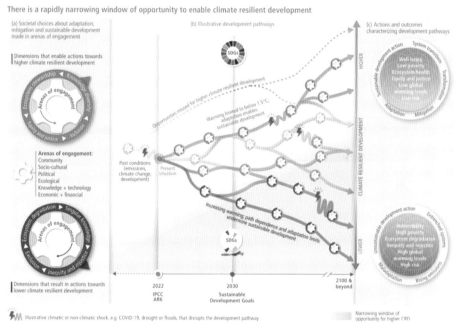

圖 3-2　氣候韌性發展路徑是成功整合溫室氣體減排與適應努力，從而支持所有人永續發展的發展軌跡。（Climate Resilient Development Pathways are development trajectories that successfully integrate GHG mitigation and adaptation efforts to support sustainable development for all.）圖片來源：IPCC。

3.3 人類影響評估（Human Impact Assessment）

目標（Intent）

確保專案開發基於對當地社區、勞動力和供應鏈社會背景的充分理解，幫助解決潛在的社會不平等問題。（To ensure that project development is guided by a thorough understanding of the social context of the local community, workforce, and supply chain, helping to address potential social inequities.）

進行人類影響評估，評估內容需包含以下類別的資訊：（Complete a human impact assessment that includes information from the following categories:）

- **人口統計**。例如，種族和族裔、性別、年齡、收入、就業率、人口密度、教育水平、家庭類型、附近弱勢群體的辨識。（Demographics. For example, race and ethnicity, gender, age, income, employment rate, population density, education levels, household types, identification of nearby vulnerable populations.）

- **當地基礎設施與土地使用**。例如，鄰近的交通和步行基礎設施、多樣化用途的鄰近設施、相關的當地或區域永續目標／承諾、適用的無障礙設施規範。（Local Infrastructure and Land Use. For example, adjacent transportation and pedestrian infrastructure, adjacent diverse uses, relevant local or regional sustainability goals/commitments, applicable accessibility code(s).）

- **人類使用與健康影響**。例如，住房的可負擔性和可用性、社會服務的可得性（例如醫療、教育、社會支持網絡）、社區安全、當地社區團體、供應鏈和施工勞動力保護。（Human Use and Health Impacts. For example, housing affordability and availability, availability of social services (e.g., healthcare, education, social support networks), community safety, local community groups, supply chain, and construction workforce protections.）

- **包含使用者體驗。**例如，自然光、視野和可操作窗戶的機會、空氣和水的環境條件、鄰近的聲景、光線和風向模式，以及周圍建築的微氣候和日照環境。（Occupant Experience. For example, opportunity for daylight, views, and operable windows, environmental conditions of air and water, adjacent soundscapes, lighting, and wind patterns within the context of surrounding buildings (microclimate, solarscape, neighboring structures).）

- **在可能的情況下，使用評估資訊來指導專案的規劃、設計、營運和維護，並描述如何考慮專案特定的策略。**（Where possible, use the information from the assessment to inform the planning, design, operations, and maintenance of the project and describe how project-specific strategies were considered.）

3.4 碳排放評估（Carbon Assessment）

目標（Intent）

了解並減少長期的直接和間接碳排放，包括現場燃燒、電網供電、冷媒和隱含碳。（To understand and reduce long-term direct and indirect carbon emissions, including on-site combustion, grid-supplied electricity, refrigerants, and embodied carbon.）

IPp：碳排放評估流程（IPp: Carbon Assessment Process）

為專案開發 25 年的碳排放預測評估。（Develop a 25-year projected carbon assessment for the project.）

評估將使用以下前提條件的數據：（The assessment will utilize the data from the following prerequisites:）

- EA 先決條件：運營碳排放預測與脫碳計畫（EA Prerequisite: Operational Carbon Projection and Decarbonization Plan）
- EA 先決條件：基礎冷媒管理（EA Prerequisite: Fundamental Refrigerant Management）
- MR 先決條件：內含碳排放的評估與量化（MR Prerequisite: Assess and Quantify Embodied Carbon）
- 可選項目：LT 得分項目：交通需求管理（Optional: LT Credit: Transportation Demand Management）

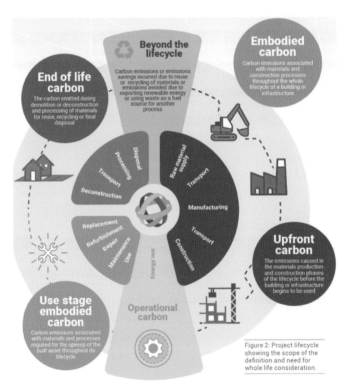

圖 3-3　建築生命週期碳評估（Building Life Cycle Carbon Assessment）。

圖片來源：Asia Pacific Embodied Carbon Primer。

　　碳排放評估（Carbon Assessment）是一個用來計算和分析一個專案或建築物在其生命週期內所產生的碳排放量的過程。它涵蓋了直接和間接碳排放，目的是了解並減少專案在營運和建設過程中的碳足跡，促進脫碳的實現。（A Carbon Assessment is a process used to calculate and analyze the carbon emissions generated by a project or building throughout its lifecycle. It covers both direct and indirect carbon emissions, with the goal of understanding and reducing the carbon footprint during the operation and construction phases, thereby promoting decarbonization.）

　　碳排放評估通常包括以下幾個方面：（A Carbon Assessment typically includes the following aspects:）

- **直接碳排放（Direct Emissions）**：這些排放來自專案現場的燃燒過程，例如天然氣、燃油等在建築物運營中的使用。（Direct Emissions: These emissions come from on-site combustion processes, such as the use of natural gas and fuel oil during building operations.）

- **間接碳排放（Indirect Emissions）**：這些排放主要來自電網供電，或其他能源間接帶來的碳排放。（Indirect Emissions: These emissions are primarily from grid-supplied electricity or other forms of energy that indirectly contribute to carbon emissions.）

- **隱含碳排放（Embodied Carbon）**：這部分碳排放與建築材料的製造、運輸、安裝、維護和廢棄有關，涵蓋了材料全生命週期的排放。（Embodied Carbon: This refers to the carbon emissions associated with the manufacturing, transportation, installation, maintenance, and disposal of building materials, covering the entire lifecycle of the materials.）

- **冷媒排放（Refrigerant Emissions）**：冷媒洩漏會對全球變暖產生顯著影響，因此冷媒管理也是碳排放評估的一部分。（Refrigerant

Emissions: Refrigerant leaks can have a significant impact on global warming, so refrigerant management is also an important part of a carbon assessment.）

碳排放評估的目標是透過這些分析，幫助專案團隊辨識高排放源並制定相應的減排策略，例如提高能源效率、使用可再生能源、優化建築材料選擇等。這一評估也是 LEED 等綠色建築標準的重要部分，推動長期的永續發展。（The goal of a Carbon Assessment is to help project teams identify major sources of emissions and develop strategies for reduction, such as improving energy efficiency, utilizing renewable energy, and optimizing the selection of building materials. This assessment is a key component of green building standards like LEED, promoting long-term sustainable development.）

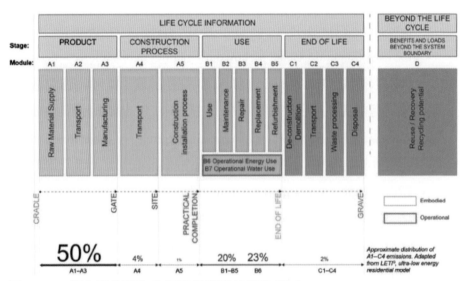

圖 3-4　一個中型住宅生命週期各階段的碳比例（Carbon Proportions Across Different Life Cycle Stages of a Medium-Sized Residential Building）。

圖片來源：英國皇家結構工程師學會（The Institution of Structural Engineers, UK）。

Decarbonization

IP Prereq	Carbon Assessment	Required
LT Credit	Compact and Connected Development	6
LT Credit	Active Travel Facilities	2
LT Credit	Transportation Demand Management	2
LT Credit	Electric Vehicles	2
SS Credit	Heat Island Reduction	2
WE Prereq	Minimum Water Efficiency	Required
WE Prereq	Enhanced Water Efficiency	6
WE Prereq	Water Reuse	2
WE Prereq	Water Metering and Leak Detection	1
EA Prereq	Operational Carbon Projection and Decarbonization Plan	Required
EA Prereq	Minimum Energy Efficiency	Required
EA Prereq	Fundamental Commissioning	Required
EA Prereq	Energy Metering and Reporting	Required
EA Prereq	Fundamental Refrigerant Management	Required
EA Credit	Electrification	5
EA Credit	Reduce Peak Thermal Loads	5
EA Credit	Enhanced Energy Efficiency	10
EA Credit	Renewable Energy	5
EA Credit	Enhanced and Ongoing Commissioning	4
EA Credit	Grid Interactive	2
EA Credit	Enhanced Refrigerant Management	2
MR Prereq	Planning for Zero Waste Operations	Required
MR Prereq	Assess Embodied Carbon	Required
MR Credit	Building and Materials Reuse	5
MR Credit	Reduce Embodied Carbon	8
MR Credit	Construction and Demolition Waste Diversion	2

圖3-5　LEED碳評估評分表範例（Example of LEED Carbon Assessment Scorecard）

3.5 整合設計流程（Integrative Design Process）

目標（Intent）

支持高性能、具成本效益的跨功能專案成果，通過早期分析和規劃系統之間的相互關聯，提供一個整體框架，使專案團隊能夠協同處理脫碳、生活品質以及生態保護與恢復，涵蓋整個 LEED 評級系統。（To support high-performance, cost-effective, and cross-functional project outcomes through early analysis and planning of the interrelationships among systems. To provide a holistic framework for project teams to collaboratively address decarbonization, quality of life, and ecosystem conservation and restoration across the entire LEED rating system.）

從前期設計開始，並持續到早期使用階段，辨識並應用機會，透過以下措施實現跨領域和建築系統的協同效應：（Beginning in pre-design and continuing throughout early occupancy, identify and apply opportunities to achieve synergies across disciplines and building systems through the following initiatives.）

- **整合團隊（Integrated Team）**：組建並召集具有多元觀點的跨學科專案團隊。通過有組織的協調確保這是一個公平的團隊協作過程。（Assemble and convene an interdisciplinary project team with diverse perspectives. Ensure the process is an equitable, team effort through organized facilitation.）

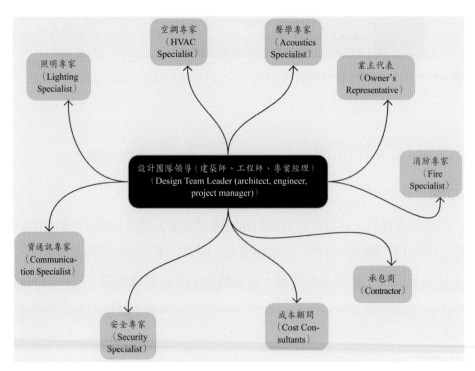

圖 3-6　整合式設計團隊（Integrative Design Team）

- **設計座談會**（**Design Charette**）：在前期設計或設計早期，與業主或業主代表以及至少四個關鍵視角的參與者進行座談會（例如：建築師、承包商、能源模擬專家、社區參與代表）。（During pre-design or early in design, conduct a charette with the owner or owner's representative and participants representing at least four key perspectives (e.g. architect, contractor, energy modeler, community engagement representatives).）

- **LEED 目標設定**（**LEED Goal Setting**）：專案團隊合作定義一組具體且可衡量的專案目標，這些目標應涵蓋 LEED v5 的影響領域：脫碳、生活品質、生態保護與恢復。將這些目標納入業主的專案要求（OPR）中。（Work as a team to define a set of specific and measurable project goals that address the LEED v5 impact areas of decarbonization, quality of life, and ecosystem conservation and restoration. Incorporate these goals into the owner's project requirements (OPR).）

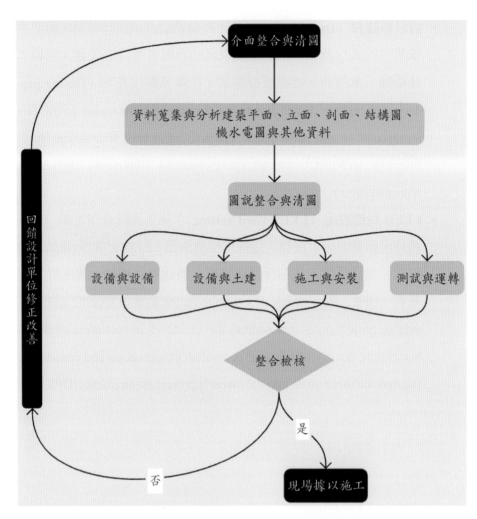

圖 3-7   整合式設計流程示意圖（Diagram of the Integrative Design Process）

第四章 基地位置和交通 （Location and Transportation）

學習目標（Learning Objectives）

- 敏感的土地保護（Protecting sensitive land）
- 高優先開發土地（Developing on high-priority sites）
- 周圍的密度和不同用途（Promoting surrounding density and diverse land use）
- 獲得高品質交通方式（Access to high-quality transit）
- 自行車設施（Providing bicycle facilities）
- 減少的停車位面積（Reducing parking lot size）
- 綠色車輛（Encouraging green vehicles）
- LEED 社區發展位置（Locating projects in LEED-certified neighborhoods）

4.1 概述（Overview）

這個得分類別的總體目標是：（The overall goals of this category are:）

- 透過適當的選址減少對環境的影響。（Reducing environmental impact through appropriate site selection.）
- 減少與交通相關的環境影響。減少因為汽車使用石化燃料導致全球變暖、酸雨、煙霧和有毒氣體排放。（Minimizing the environmental impacts related to transportation, such as reducing global warming, acid rain, smog, and toxic gas emissions caused by vehicles using fossil fuels.）

- 透過增加運動與活動提高大眾健康。（Enhancing public health by promoting physical activity and movement.）

基地選擇對於綠建築設計規劃的影響，可由下列2個主要項目來討論：（The impact of site selection on green building design can be discussed through the following two key aspects:）

基地選擇（**Site Selection**）

- LEED 對於鄰里發展的位置（LEED for Neighborhood Development Location）
- 環境敏感地保護（Sensitive Land Protection）
- 優先開發基地（High Priority Site）
- 周圍密度與不同的用途（Surrounding Density and Diverse Uses）

交通（**Transportation**）

- 高品質的運輸（Access to Quality Transit）
- 自行車設施（Bicycle Facilities）
- 減少停車（Reduced Parking）
- 減少停車範圍（Reduced Parking Footprint）
- 環保車輛（Green Vehicles）

基地選擇（Site Selection）

- 選擇已取得 **LEED** 認證的建築物（**Choose a LEED Certified Building**）：對於室內設計專案，選擇那些沒有建造新建築物的專案計畫，可以選擇一個LEED-CS和LEED-NC認證的建築來認證。（For interior projects, choose projects that are not constructing a new building, can choose to locate in a LEED-CS or LEED-NC Certified building.）

inhabitat.com　　　　　　　www.aiatopten.org

圖 4-1　已取得 LEED 認證之建築物（白金級）（Genzyme Center, Cambridge, Massachusetts (LEED Platinum)）

- 選擇取得 **LEED ND** 認證的基地（**Locate in a certified LEED ND site**）：整個建築專案位於 LEED 認證的邊界內發展之專案計畫。（For whole building projects, locate the project within the boundary of a development under LEED for ND.）

智慧型成長（Smart Growth）

　　專案中若申請 LEED ND，則可以涵蓋位置與交通 LT 中所有的得分項目，並有助於智慧成長（**Smart Growth**）完整開發的最佳策略。智慧

成長是城市規劃和交通的理論，在城市中就能步行移動各建築，避免城市擴張，簡單來說，即是將開發範圍集中。（If a project applies for LEED Neighborhood Development (LEED ND) certification, it can cover all the credit items under the Location and Transportation (LT) category. This contributes to the best strategies for Smart Growth, which is an urban planning and transportation theory. Smart Growth focuses on creating walkable urban environments, reducing the need for car travel, and preventing urban sprawl by concentrating development in specific areas. In essence, it encourages compact, sustainable urban development that promotes efficient land use and enhances quality of life.）

　　智慧成長是用一種更好的方式來建立和維護我們的城鎮和城市。意味著建設城市，郊區和農村社區成爲住房使用，並且將交通選擇靠近工作、商店和學校的地方，以增近方便。（Smart growth is an urban planning and transportation theory that concerntrates growth in compact walkable urban centres to avoid sprawl.）

　　這種方法可以支持當地經濟和保護環境。且土地利用應住商混合、交通導向、行人優先，並對單車友善。著重永續性多於短期利益。（It also advocates compact, transit-oriented, walkable, bicycle-friendly land use, including neighborhood schools, complete streets and mix-used development with a range of housing choices.）

　　其原則包含以下幾項：

圖 4-2 智慧成長

　　智慧型成長是一個具有保護開放空間與綠地、社區再活化、社區多樣性、房價合理以及較多交通選擇性的完善規劃。美國希望朝向這樣的發展目標發展，而台灣目前其實開發的模式已經接近於所謂的智慧型成長。智慧成長包含了以下的十個原則：（Smart Growth is an urban planning strategy that emphasizes the protection of open spaces and green areas, the revitalization of communities, the promotion of diversity within communities, affordable housing, and offering a wider range of transportation options. The U.S. has set these principles as development goals, and Taiwan's current development model is already quite close to the ideals of Smart Growth. Smart Growth includes the following ten principles:）

1. 混合式開發（Mix land uses）
2. 利用緊湊／高密度的建築設計（Take advantage of compact building design）

3. 提供住宅房型多樣性（Create a range of housing opportunities and choices）

4. 建立適合徒步的社區（Create walkable neighborhoods）

5. 建立社區特色，加強居民與彼此以及社區的連結（Foster distinctive, attractive communities with a strong sense of place）

6. 保持開放空間、農田、自然之美和關鍵環境領域（Preserve open space, farmland, natural beauty, and critical environmental areas）

7. 加強對現有的社區直接開發（Strengthen and direct development towards existing communities）

8. 提供多元交通方案（Provide a variety of transportation choices）

9. 可預測的、公平且符合成本效益的開發決策（Make development decisions predictable, fair, and cost effective）

10. 鼓勵社區和利益相關者合作討論開發決策（Encourage community and stakeholder collaboration in development decisions）

4.2 環境敏感地保護（Sensitive Land Protection）

保護棲息物種（Protect Habitat）

開發地不該選擇生態敏感地區。（Give preference to location that do not include sensitive site elements and land type.）

- 自然（Natural）
- 鄉村（Rural）
- 郊區（Sub-Urban）
- 都市（General Urban）

- 市區（Town Center）
- 市中心（Town Core）

圖 4-3　開發區域示意

　　本項的目的是為了避免開發環境敏感的地點，因此減少了開發造成的相關環境影響。（The intent of this credit is to avoid development of enviromentally sensitives sites and hence reduces the impacts with the development.）

1. 實現這一得分，最好的策略是開發一個以前開發的土地。以前開發的土地都具有現有的電力網絡、供水、汙水處理基礎設施建設，等於是降低了對基礎設施之要求的負荷。（Achieve this credit is to develope a previously developed site have existing infrastructure like electricity network, water, sewage, telecom.）

2. 不要開發環境敏感地點。（Do not develope environmentally sensitive sites.）以下例子被認為是環境敏感地區：

圖 4-4　濕地、湖泊、自然公園、河川、物種棲息地

瀕危物種：處於高風險成為絕種的動植物。（Endangered species: Population of organism which is at risk of becoming extinct.）

濕地：濕地是有著高土壤水分的土地，無論是永久性或季節性濕地。（Wetland: Is an area of land whose soil is saturated with moisture either permanently or seasonally.）

環境敏感地保護（Sensitive Land Protection）

避免在不適當的基地上進行開發，並藉由建築物在基地上的配置方式達到減少對周邊環境的影響。（Avoid the development of environmentally sensitive lands and reduce the environmental impact from the location of a building on a site.）

banella.com

圖 4-5　環境敏感之地區

　　在 LEED 之中提供了七項避免在下列土地上進行建築物、景觀設施、道路或停車場等之開發：（In the LEED (Leadership in Energy and Environmental Design) system, there are seven specific types of land where development of buildings, landscaping, roads, or parking areas should be avoided. These restrictions aim to protect sensitive areas and promote sustainable land use. Here are the types of land where development should be prevented:）

1. 未開發素地、綠地（**Greenfield sites**）

2. 優良農耕地（Prime **farmland** indentified by **Natural Resources Conservation Service (NRCS)**）

3. 不管在任何時間，都避免在容易受到洪水或淹水侵害的地區開發（洪汜區）（Areas below **Floodplains** designated by the **Federal Emergency Management agency (FEMA)**）

4. 公園用地（Public parkland）

5. 受威脅或瀕臨絕種生物的棲息地，土地具有確定爲瀕危的物種（habitat for **threatened or endangered species** determined by **The Natural Heritage Program** or the **state wildlife agency**）

6. 距離自然水體 50 呎以內範圍之土地（Within 50 feet of **wetlands**）：
   這邊指的是自然水體，人造水體則不算在內。

7. 距離濕地 100 呎以內範圍之土地（Within 100 feet of a **water body**）

4.3 公平發展（Equitable Development）

這一得分項目希望：

- 鼓勵專案開發於已經開發的位置。（Encourge project location in areas with development constraints.）

- 減少對綠地和農田的開發負荷。（Reduce the load on Greenfield sites and farm lands.）

策略：（Strategy）

1. 將基地設在現有的已開發基地。（Locate the project in infill site of an existing development.）

2. 在都市中找尋還可以建設之空地。（Infill development involves development with in urban area.）

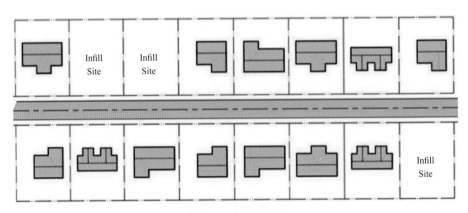

圖 4-6　密集開發

選擇優先開發基地（Choose High-Priority Site）

建立在以前開發或受汙染基地上。舊的建築物通常有很好的交通位置，可獲得現有的公共服務系統及公共交通工具。（Build on a **previously developed** or **brownfield** site. Old buildings are often well located, with access to existing services and public transportation.）

圖 4-7　受汙染或高密度地區

選擇已開發基地（Use Infill Development）

建立在先前開發過的地點，通常意味著有現有的基礎設施，例如道路、水電等其他服務。在歷史區重新開發，也可以適應性再利用並減少城市擴張。由於先前開發的地區已經將土地使用破壞，所以後來的建築物若可以開發在之前已經開發過的土地，則對於環境是較好的。（Building on previously developed site usually means there is existing infrastructure like roads, utilities and other services. The redevelopment of sites in **historic districts** can also reduce urban sprawl through **adaptive reuse.**）

也鼓勵將建築物開發在兩棟建築物中間的空間（A site in-between existing structures LEED），如下圖所示，既有兩棟建物中若開發新建築是優良的。

rock.gu.wordpress.com

圖 4-8　填塞式開發

選擇受汙染之褐地（Redevelop Brownfield Sites）

　　在地方、州或聯邦政府機關所認定之受汙染基地（褐地）上進行開發，針對土壤及地下水已確認受汙染之褐地，並依主管機關要求執行必要之整治。（Locate on a brownfield where soil or groundwater contamination has been identified, and where the local, state,or national authority requires its remediation.）

圖 4-9　褐地整治

重建褐（棕）地

　　褐（棕）地是指在城市規劃用於先前在於工業用途或一些商業用途之土地的地方。這樣的土地可能被汙染的危險廢棄物汙染。一旦清理處理好，這樣的區域可以成爲主要的開發場所，如公園。而土地更嚴重汙染，並具有高濃度的危險廢物和汙染，並不會被歸類在褐地。（Brownfield sites are contaminated or perceived to be contaminated sites. Brownfield sites requites remediation prior to development.）

　　褐地是受到汙染的場址。褐地需要先行處理再開發。一個土地可能在以下情況可以獲得褐地得分：

- 由政府機構網站聲明爲褐地（Site declared as brown field by government agencies.）
- 如果該土地是由當地的自願清理計畫修復（If the site is remediated by local voluntary programs.）
- 汙染物是按照美國 ASTM 標準進行環境現場評估過程中發現的（Contaminants are identified during environmental site assessment performed as per ASTM standards.）

圖 4-10　褐（棕）地

優先開發地區（Priority Destinations）

　　優先開發地區指的是由政府單位鼓勵或支持開發的地區，其中大部分是為了鼓勵投資在經濟較貧困與高失業率地區或低收入戶社區。（Priority destinations are the sites whose development is encouraged or supported by Government. Most of which are intended to encourage investment in economically disadvantage or low-income areas.）

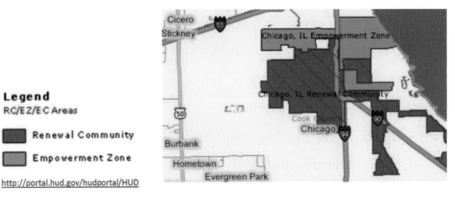

圖 4-11　優先開發之地區

　　指定優先區域：優先開發基地是其開發受到政府鼓勵或支持的地區。（Designated Priority Area: A priority development site is an area where development is encouraged or supported by the government.）

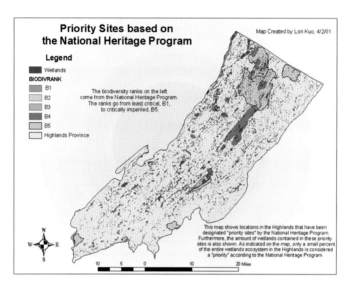

圖 4-12 優先指定地圖

來源：www.deathstar.rutgers.edu.

4.4 緊湊與連接式發展（Compact and Connected Development）

本得分項目的目的是：

- 促進現有基礎設施等區域的開發（To promote development in areas with existing infrastructure）
- 促進步行能力與運輸效率（Promote walk-ability and transportation efficiency）
- 降低車輛行駛距離（Reduce vehicle distance traveled）
- 透過鼓勵日常體力活動提高公眾健康（Improve public health by daily physical activity）

LEED 鼓勵高密度的開發。土地是寶貴的資源；高密度發展具有較小

的開發面積建設，從而保護不去開發全新土地和農田。分數分配根據專案內 1/4 英里的密度和周圍環境來評估。(LEED encourages high dense development. Land is a precious resource; high dense development has lesser building footprint thereby protecting greenfield sites and farmlands from development. Points allotted based on density of project and surroundings within 1/4 mile.)

圖 4-13　高密度開發

LEED 鼓勵開發具有多元的用途使用。(LEED encourages development with diverse use.) 一個多樣化的使用地點應具有每天的日常需求場所，像超市、餐廳、洗衣店、醫院、步行距離之內的銀行等。這樣可以減少汽車的使用，提高居住者的身體活動。分數基於不同使用場所的數量，從專案出入口內半英里的步行距離計算來給予。計算時從基地內畫出一個圓形，涵蓋的範圍之中含有多少服務的機構。(A diverse use location shall have all day-to-day requirements like super market, restaurant, laundry, hospital, bank etc within walking distance. This can reduce automobile usage and increase physical activity of the occupants. Points are allotted based on number of diverse use spaces within 1/2 mile walking distance from project entrance.)

在密度高的地方開發建設（Develop in Dense Area）

將基地設於高密度的地方，例如市中心。（Locate on a site whose surrounding areas have higher density, such as a city center.）

優點（Advantages）

- 充足的公共建設（Sufficient infrastructure）
- 限制城市擴張（Limits urban sprawl）
- 通常有公共交通路線（Usually on public transportation lines）
- 減少暴雨逕流（Generate less storm water runoff）
- 周圍建築密度（Surrounding Density）：周邊密度由密度半徑內的所有建築物的平均平方英尺測量。（The Surrounding density is measured by the average square footage of all buildings within the density radius.）
- 密度半徑（Density radius）：密度半徑計算是在位置周圍繪製一個圓，確定所有屬性並在半徑內確認密度。（The density radius calculation is used to draw a circle around the site and determine all properties within the radius to determine the density.）
- 增加開發密度（Increase Density）：創建一個更小的開發範圍，將容積率最大化，考慮向上建設取代往外擴張。（Create a smaller footprint and maximize the floor-area ratio, consider building up, instead of out.）

圖 4-14　畫一半徑計算周圍密度

圖 4-15　高密度之城市地區

增加多樣化服務設施（Increase Diversity of Uses）

多樣化服務設施是人們定期會使用的服務，這些必須是行人可以步行前往而不會被阻隔。（**Diverse uses** are common service that people might use regularly. There must be **pedestrian access** to these services without being blocked.）

不同用途包含（Diverse Use Includes）

(1) 銀行；(2) 禮拜堂；(3) 便利商店；(4) 日間托兒所；(5) 乾洗店；(6) 消防隊；(7) 美容院；(8) 五金行；(9) 自助洗衣店；(10) 圖書館；(11) 醫療／牙醫診所；(12) 老人養護設施；(13) 公園；(14) 藥房；(15) 郵局；(16) 餐廳；(17) 學校；(18) 超市；(19) 電影院；(20) 社區中心；(21) 健身房；(22) 博物館等。（(1) Bank; (2) Chapel; (3) Convenience Store; (4) Daycare Center; (5) Dry Cleaner; (6) Fire Station; (7) Hair Salon; (8) Hardware Store; (9) Laundromat; (10) Library; (11) Medical/Dental Office; (12) Nursing Home; (13) Park; (14) Pharmacy; (15) Post Office; (16) Restaurant; (17) School; (18) Supermarket; (19) Theater; (20) Community Center; (21) Gym; (22) Museum, etc.）

NOT Including（不包含）：

可移動的服務設施：ATM；販賣機等。（Portable Service Facilities: ATM, vending machines, etc.）

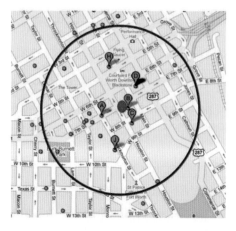

圖 4-16　位於基地周圍
　　　　之服務設施

步行距離超出範
圍，不得計入

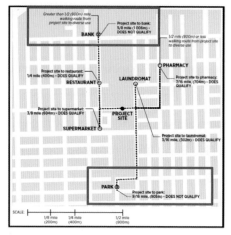

步行距離超出範
圍，不得計入

圖 4-17　周圍服務設施以步行距離計算

來源：USGBC reference guide.

交通（Transportation）

2007 年，交通運輸佔全美國溫室氣體排放量的 32%。（In 2007, Transportation accounts for 32% of the nation's greenhouse gas emissions.）

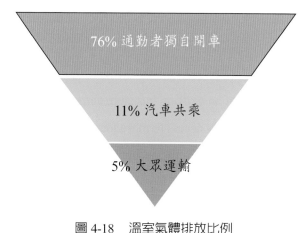

圖 4-18　溫室氣體排放比例

4.5 交通需求管理（Transportation Demand Management）

得分項目的目的：（The purpose of LEED credits）

- 鼓勵發展公共交通設施。（Encourage development with transportation facilities.）

- 減少溫室氣體排放、空氣汙染以及與汽車使用相關的其他環境和公眾健康危害。（Reduce greenhouse gas emissions, air pollution, and other environmental and public health harms associated with automobile use.）

- 選擇的基地有捷運或巴士站經過。（A site is located near a mass transit or bus station.）

- 如果該基地在行人步行距離半英里內有永久、現成的火車站／地鐵站／巴士轉運站，則會提升交通的便利性。（The site is within a half-mile (800 meters) walking distance of a permanent, operational train station, subway station, or bus transfer station. This boosts transportation convenience by offering nearby public transit options.）

- 如果基地內在行人步行距離 1/4 英里內就可以到達現有的公車站、共乘車或小型巴士。（The site is within a quarter-mile (400 meters) walking distance of an existing bus stop, carpool location, or shuttle service. This ensures easy access to more localized transit services, improving overall connectivity for the occupants of the building.）

策略：（Strategies）

　　設於大眾運輸工具附近（Locate the project near mass transit）：選擇一個現有的交通運輸網路。步行距離可以鼓勵居住者使用接駁車輛。（Select a project site in an area served by an existing transportation network. Walking distances can encourage the building occupants to access.）

圖 4-19　公共交通工具

定位建築物的步行距離：（Locate Building with Walking Distance to）

- 公車（Public Bus）
- 校車（Campus Bus）
- 私人接駁車（Private Shuttle）

- 通勤鐵路（Commuter rail）
- 輕軌（Light rail）
- 地鐵站（Subway station）
- 通勤渡輪站（Commuter ferry stations）

必須同時滿足工作日和週末之旅最低的轉運服務。（Must meet both weekday and weekend trip minimums for the transit services.）

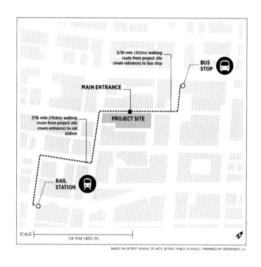

圖 4-20　基地與大眾交通之步行距離

來源：USGBC reference guide.

自行車相關設施（Bicycle Facilities）

得分項目的目的：（The purpose of LEED credits:）

- 促進自行車和運輸效率（Promote bicycling and transportation efficiency）
- 降低車輛行駛距離（Reduce vehicle distance traveled）

- 透過增加體力活動提高公眾健康（Improve public health by increased physical activity）

策略：（Strategies:）

- 選擇的基地裡面有自行車道（Select site with has bicycle track）
- 在住宅建築中，提供安全的自行車停車位。（Provide safety bicycle racks in residential building.）
- 在商業大樓中，提供安全的自行車停車位、淋浴間和更衣室。（Provide bicycle racks, sbower rooms and dressing rooms in commercial building.）

圖 4-21　自行車停車的地方

騎腳踏車的優點（Bicycling Benefits）

　　騎自行車提供了許多個人和全球的利益。每騎一英里減少了近 1 磅的二氧化碳排放量，並透過鼓勵身體活動改善大眾健康。（Bicycling offers many individual and global benefits. For every mile pedaled rather than driven, nearly 1 pound of carbon dioxide emissions is avoided, as well as improving public health by encouraging physical activity.）

每 2000 個旅客所需之停車空間

汽車	公車	單車
7 個足球場	半個足球場	1/8 個足球場

圖 4-22　停車空間之比較

自行車專用道路網（Bicycle Network）

　　選擇地點是靠近自行車專用道（道路、小徑、指定的自行車道以及低速公路），並提供安全自行車架和淋浴設施，給居民騎自行車上班的選項。（Select project sites within proximity to bicycle networks (paths, trails, designated bike lanes, and slow-speed roadways), and provide **secure bike racks** and **shower facilities** to give occupants the option of biking to work.）

圖 4-23　自行車道與停車空間

提供腳踏車架以及更衣室 / 淋浴間（Bicycle Storage & Changing Rooms）

非住宅區：（For None Residential）

- 短期自行車架給 5% 訪客（Short-term Bicycle racks for 5% Peak Visitors）
- 長期自行車架給 2.5% 常態建築使用人員（Long-term Bicycle racks for 2.5% regular building occupants）
- 提供淋浴間給常態建築使用人員（Provide shower rooms（w/ locker）for regular building occupants）

住宅區：（Residential）

- 短期自行車架給 5% 訪客（Short-term Bicycle racks for 2.5% Peak Visitors）
- 長期自行車架給 30% 常態建築使用人員（Long-term Bicycle racks for 30% regular building occupants）

常態建築使用人員		訪客
• 員工（Employee） • 職員（Staff） • 義工 • 居民	• 小學生及中學生 • 旅館房客 • 住院病人	• 零售消費者（客戶） • 門診病人 • 尖峰門診病人 • 義工 • 高等教育機構學生：如大學

減少停車場開發面積（Reduced Parking Footprint）

得分項目的目的是盡量減少與停車相關的設施，可以減少對環境的影響（This credit is to minimize the environmental impact associated with parking facilities.）包括．

- 汽車的依賴（Automobile dependence）
- 土地消耗（Land consumption）
- 雨水徑流（Rainwater off）

策略：（Strategies:）

- 根據當地區域限制的要求來限制本地停車位的數量。（Limit the number of parking to local zoning requirements.）
- 如果當地並沒有區域限制停車數量的要求，則可以用交通運輸工程手冊（Institute of Transportation Engineering Handbook）來做參考。

名詞小教室

　　地方分區管理局（Local Zoning Authority）：是一個機構，它制定的開放空間的要求、建築高度、停車需求和建築物的安全要求。在許多城市被用來做為當地管理標準。（**Local Zoning Authority:** Agency which regulates open space requirement, building height, parking requirement and safety requirements of building. In many cities, municipality acts as local zoning authority.）

提供獎勵（Offer Incentives）

　　對居民的一種（減少停車場開發）的通勤替代激勵計畫，例如「金錢」。（Develop an alternative commuting incentive program for building occupants, such as「Cash」rewards.）

- 鼓勵使用替代交通工具（Support alternative transportation）推廣單人通勤碳排高通勤方式，例如公告版上的共乘團體或公車時間。

　　（Promote alternatives to single-occupancy car commuting at the building, such as a bulletin board with carpool groups and bus schedules）

圖 4-24　減少單獨駕車之宣導廣告

縮小停車場面積（Minimize Parking）

- 製造一個小型停車區域給居民，因停車位有限，將會減少汽車使用。（Create the smallest parking area for occupant's need. If parking is limited, there might be less automobile use.）

- 汽車車位設置不得超過法定數量。（Do not exceed the minimum parking requirement by local zoning code.）

- 汽車車位設置數量需低於美國 ITE Transportation Planning Handbook 中建議之基準值。（Without zoning requirement: Use the ITE (Institute of Transportation Engineering) Transportation Planning Handbook.）

圖 4-25　汽車共乘專用停車位

限制停車數量（Limit Parking Capacity）

- 與鄰近的建築共用停車場（Consider sharing parking lots with neighboring properties）
- 不提供新的停車位（Provide no new parking）
- 保留專用停車位（Reserve preferred parking spaces for Carpool & vanpool (5% total parking)）
- 提供交通共乘服務（Provide Car-share services）

圖 4-26　汽車共乘計畫之示意圖與應用程式

4.6 電動車（Electric Vehicles）

透過鼓勵替代傳統燃料的汽車減少汙染。（Encouraging the use of alternative fuel vehicles helps reduce pollution by replacing traditional gasoline and diesel-powered cars with more environmentally friendly options.）

策略：（Strategies:）

安裝符合列出門檻的電動車供電設備（EVSE）。（Install electric vehicle supply equipment (EVSE) meeting the thresholds listed）

EVSE 必須符合以下標準：（EVSE must meet the following criteria:）

- 提供符合製造商要求和《全國電氣規範》（NFPA 70）要求的 Level 2 或 Level 3 充電能力。（Provide Level 2 or Level 3 charging capacity per the manufacturer's requirements and the requirements of the National Electrical Code (NFPA 70).）

- 每個指定空間的電壓為 208～240 伏特或更高。（208-240 volts or greater for each required space.）

- 符合 ENERGY STAR 認證的 EVSE 連接功能標準，並能夠響應分時市場信號（例如價格）。（Meet the connected functionality criteria for ENERGY STAR certified EVSE and be capable of responding to time-of-use market signals (e.g. price).）

- 至少一個 EV 充電站必須是無障礙停車位，寬至少 9 英尺（2.5 公尺），並配有 5 英尺（1.5 公尺）寬的通道，且具備為行動不便、行走困難和視力受限人群提供便利的充電站無障礙設施。（At least one EV charging station must be an accessible parking space at least 9-feet (2.5-meters) wide with a 5-feet (1.5-meters) access aisle and have charging station accessibility features for use by persons with mobility, ambulatory and visual limitations.）

　　綠色車輛、對環境友好的車輛，或使用某些替代燃料行駛的交通工具，比用汽油或柴油產生較少的有害影響。目前，這個詞在一些國家用於任何車輛符合或超過了更嚴格的歐洲排放標準（如 EURO6），或加州的零排放車輛標準（如 ZEV、ULEV、SULEV、PZEV），或在幾個國家之中的低碳燃料標準。（Green vehicles, or environmentally friendly vehicles, are those that have a reduced environmental impact compared to traditional gasoline or diesel vehicles. These vehicles typically run on alternative fuels and produce fewer harmful emissions. In some countries, the term "green vehicles" refers to those that meet or exceed stringent emission standards, such as:EURO 6: A European emission standard for vehicles that limits nitrogen oxides (NOx) and particulate matter.California's Zero Emission Vehicle (ZEV) standards: Includes vehicles like ZEVs, Ultra-Low Emission Vehicles (ULEVs), Super Ultra-Low Emission Vehicles (SULEVs), and Partial Zero Emission Vehicles (PZEVs).Low Carbon Fuel Standards: Implemented in several countries, these standards encourage the use of fuels that emit fewer greenhouse gases.）

圖 4-27　LEV/FEV 優先停車位

EV-ready 停車位必須提供完整的電路安裝，包括 208/240 伏特、40 安培配電盤容量，以及終止於接線盒或充電插座的導管（線槽）和配線。（EV-ready parking spaces must provide a full circuit installation including 208/240V, 40-amp panel capacity, and conduit (raceway) with wiring that terminates in a junction box or charging outlet.）

在選項 1 中安裝了 EVSE 並計入得分的任何車位，不得計入選項 2 中的 EV-ready 車位得分。（Any space with an installed EVSE counted for credit under Option 1 may not be counted for credit as an EV-ready space under Option 2.）

名詞小教室

- 低排放車輛（Low emitting vehicles）是被列為零排放汽車（ZEV），符合美國加州空氣資源委員會（CARB）的車輛。(**Low emitting vehicles** are vehicles that are classified as zero emission vehicles (ZEV) by California air resource board (CARB).）

- 高效率的汽車（Fuel efficient vehicles）是已經在美國節能經濟委員會（American Council for an Energy Efficient Economy, ACEEE）年度汽車評級指南得到綠色得分最低分數 45 以上的車輛。(**Fuel efficient vehicles** are vehicles which have achieved a minimum green score of 45 on the American Council for an Energy Efficient Economy (ACEEE) annual vehicle rating guide.）

代替燃料汽車的環保汽車可能是電動汽車，或搭載氫、乙醇、天然氣或生物燃料汽車。汽車有資格被列為：（Green Vehicles Alternative fuel vehicles may be electric cars, or cars powered by hydrogen, ethanol, natural gas, or biofuel. Cars that are eligible are classified as:）

- 低汙染環保車輛（Low Emitting Vehicles (LEV): **Zero Emission Vehicles (ZEVs)** by the California Air Resources Board (CARB)）

- 高燃油效率環保車輛（**Fuel Efficient Vehicles** (FEV): Cars that have earned a **Green Score of 45** or more from the American Council for an Energy Efficient Economy (ACEEE)）

- 鼓勵使用環保車（Promote alternative-fuel vehicles）

- 提供優先的停車位或是有便利的燃料補給站（Provide best parking spots, or a convenient refueling station on the site.）

  ■ 保留專用停車位（Reserve preferred parking spaces for FEV）

■ 提供環保車停車優惠（Provide incentives/parking discounts for FEV drivers）

■ 提供環保車（Provide FEV vehicles）

■ 設置電動車充電站（Install electrical vehicle supply equipment（EVSE））

油電混合車　　　　電能車　　　　　瓦斯車　　　　氫燃料車

圖 4-28　綠能汽車

名詞解釋（Glossary）

- 替代能源交通工具（Alternative-fuel vehicles）

使用低汙染、無汽油燃料，如電、氫氣、丙烷或壓縮天然氣、液體天然氣、甲醇和乙醇的車輛。（Vehicles that use low-polluting, non-gasoline fuels, such as electricity, hydrogen, propane or compressed natural gas, liquid natural gas, methanol, and ethanol.）

有效的油 - 電混合車輛包括在該組 LEED 的目的。（Efficient gas-electric hybrid vehicles are included in this group for LEED purposes.）

- 高效率交通工具（Fuel-efficient vehicles）

車輛在美國委員會的能源效率經濟年度車評導向中有達到最低綠色環保分數 45 分。（Vehicles that have achieved a minimum green score of 45 on the American Council for an Energy Efficient Economy annual vehicle-rating guide.）

- 低碳排放交通工具（Low-emitting vehicles）

  由美國加州空氣資源委員會將車輛列為零排放汽車（ZEVS）。（Vehicles classified as zero-emission vehicles (ZEVs) by the California Air Resources Board.）

題目（Practice Questions）

1. 什麼指標是與建築專案相關的交通影響的最佳指標？（What metric is the best indicator of transportation impacts associated with a building project?）

   a. 街道網格線密度（Street grid density）

   b. 可用性公共交通（Availability of public transportation）

   c. 車輛行駛里程（Vehicle miles traveled）

   d. 停車場容量（Parking capacity）

   車輛行駛里程是用來估計與計劃運輸行駛英里的距離。（Vehicle miles traveled is a measure of transportation demand that estimates the travel miles associated with a project.）

2. 如何使遠離公共交通的基地減少其交通運輸對環境的影響？（How could a remote project located away from public transportation reduce its transportation effects?）

   a. 基地附近有公園（Locate a project site near a park）

   b. 為用戶錯開工作時間（Stagger work hours for users）

   c. 安裝透水路面（Install pervious pavement）

   d. 鼓勵合夥搭車（Encourage carpooling）

   交通運輸需求的策略，可以減少單獨駕駛使用包括：（Transportation demand strategies that can reduce single-occupancy vehicle use include:）

   - 在基地附近有公共交通（Locating a project near public transportation）

- 鼓勵合夥搭車（Encouraging carpooling）

- 鼓勵走路或騎腳踏車（Encouraging walking or bicycling）

- 提供綠色車輛優先停車（Providing preferred parking for green vehicles）

- 對公車、鐵路及渡輪提供折扣的交通通行證（Discounted transportation passes for buses, rails, ferry's, etc.）

- 電子通勤（Telecommuting）

- 壓縮工作週（Compressed workweeks）

3. 下列何者不是智慧成長的例子？（What is NOT an example of smart growth?）

a. 住宅開發在以前開發的站點，靠近商店和學校（Residential development on a previously developed site located near shops and schools）

b. 零售、辦公和住宅排屋的基地在一個加油站前（Retail, Office and residential townhouses on the site of a former gas station）

c. 鄰里設置辦公室和商店的地點於步行距離之內有公共交通（Neighborhood design that has offices and shops within walking distance to public transportation）

d. 在一個遠離現有的發展和基礎設施的地方開發基地（Development of a site that is far from existing development and infrastructure）

智慧型成長必須開發在靠近交通、住房和就職區，避免開發在空地和農田。（Smart growth is developing in areas near transportation, housing, and jobs in order to leave open spaces and farmland free from development.）

本發展的實施例子最有可能是在一個不適合發展的未開發地區（This example of development is most likely on a greenfield which is not preferable to infill development.）

4. 專案團隊能使用下列何者來識別一個敏感的棲息地？（Which of the fol-

lowing can project teams use to identify a sensitive habitat?）

a. 自然遺產計畫（**The Natural Heritage Program**）

b. 美國聯邦法規法典（The U.S. Code of Federal Regulations）

c. 國家魚類和野生動物機構（或當地同級機構）（**State fish and wildlife agencies (or local equivalent)**）

d. 林業發展的國際協會（ISA）（The international Society of Arboriculture (ISA)）

專案團隊在美國應先與國家自然遺產計畫和國家野生動物機構聯結，以確保開發地不會威脅到任何生物棲息地或瀕危的物種。這些機構的人員也可以幫助確定陸地上的敏感棲息地。（Project teams in the U.S. should contact the state Natual Heritage Program and state wildlife agency to determine whether any habitat for threatened or endangered species has been or is likely to be found on the project site. People from these agencies can assist with determining sensitive habitats on land.）

5. 開發地點位於步行距離內可減低：（Development located within walking distance of diverse uses reduce:）

a. 附近的水體沉積作用（Sedimentation of nearby water bodies）

b. 溫室氣體排放（**Greenhouse gas emissions**）

c. 空氣汙染（**Air pollution**）

d. 用水（Water use）

第五章　永續基地（Sustainable Sites）

學習目標（Learning Objectives）

- 建築施工汙染防治（Construction Pollution Prevention）
- 基地評估（Site assessment）
- 保護或恢復生物棲息地（Site development-protect or restore habitat）
- 基地開發保持開放空間（Open space）
- 雨水逕流管理（Rainwater management）
- 熱島效應（Heat island effect）
- 減少光汙染（Light pollution reduction）

5.1 最小化基地干擾（Minimize Site Disturbance）

為了減少施工活動的汙染，主要包括施工造成的侵蝕、沉積與空氣中的灰塵生成。（To reduce pollution from construction activities, including erosion, sedimentation.）

- 因為侵蝕造成表土的流失。（Erosion results in loss of topsoil.）
  頂部的土壤中含有豐富營養成分的生物和有機物質，它可以幫助植物生長。（Top soil is rich in biological and organic matter, support plant life.）

- 表土的流失降低了基地種植的能力。因為景觀的需求這可能會進一步增加化肥，從而對環境造成更多的傷害。（Loss of topsoil reduces the capacity of the site to support plantation. This may further increase the fertilizer requirement for the landscape thereby making more damage to the environment.）

- 豐富的雨水流經施工現場會因為建築材料而導致水質汙染。（Storm water run-off from the construction site is rich in contaminants because of construction materials and causes water pollution.）

- 由於侵蝕，風汙染空氣與顆粒物和懸浮顆粒會導致人類呼吸問題。（Erosion due to wind pollutes the air with particulate matter and suspended particles which results in respiratory problems for human.）

專案團隊應該制定並實施全面的侵蝕和沉積控制（ESC）計畫與 (1)2012 年美國環保局的施工許可證、(2) 國家汙染物排放消除計畫或 (3) 適用當地標準三者之中以嚴格為準。（Project team should prepare and implement a comprehensive Erosion and Sedimentation Cotrol (ESC) plan complying with *EPA 2012 Construction General Permit, National Pollutant Discharge Elimination Scheme or applicable local standards* whichever is stringent. Following are the potential strategies the project team can incorporate.）

其他的侵蝕和沉積控制（ESC）的策略可分為穩定的方式和結構的方式。（Other erosion and sedimentation control (ESC) strategies can be categorized into two main types:）

- **臨時植生**：種植快速生長的草本植物暫時穩定土壤。（Temporary Seeding: Plant fast growing grasses to temporarily stabilize the soil.）

- **永久植生**：種植草地、樹木和灌木以永久穩定土壤。（Permanent Seeding: Plant grass, trees and shrubs to permanently stabilize the soil.）

- **覆蓋**：地膜覆蓋試驗過程的傳播材料，像木屑、秸稈、乾草、草、木屑或礫石在表層土壤穩定它。

降低施工衝擊（Minimize Construction Impacts）

　　在施工之前，必須有最佳管理措施計畫，以解決施工活動的汙染及防止侵蝕和沉積。（Prior to construction, there must be a plan to implement best management practices to address construction activities and prevent erosion and sedimentation.）

暫時性植生穩定　　　　　　木屑覆蓋　　　　　　沉砂池
（Temporary Seeding）　　（Mulching）　　（Sedimentation Trap）

圖 5-1　管理措施計畫

　　制定並實施「侵蝕與泥砂控制計畫」以降低施工過程所產生的汙染。（Create and implement an Erosion and Sedimentation Control Plan (ESC Plan) to reduce construction pollution.）

　　(1)防止暴雨造成土壤侵蝕（Prevents loss of soil by stormwater runoff and wind）

　　(2)防止水道泥砂沉積（Prevents sedimentation of storm sewer and receiving streams）

　　(3)防止灰塵造成空氣汙染（Prevents pollution of the air with dust and particulate matter）

相關規範（Standards）：

　　2012 美國環保署一般施工許可或當地法規。

　　（2012 EPA General Construction Permit or local codes.）

基地評估（Site Assessment）

1. 策略（Strategies）

基地評估基地情況在設計初期，藉由基地評估可以評估綠色計畫，有關基地設計相關決定之前，基地評估是整體過程的一部分。（A site assessment assesses site conditions before design to evaluate sustainable options and inform related decisions about site design. A site assessment is part of the **integrative process**.）

基地評估檢討：（An assessment reviews the sites:）

- 地形（Topography）
- 水文（Hydrology）
- 氣候（Climate）
- 植栽（Vegetation）
- 土壤（Soils）
- 人為活動（Human use）
- 健康影響（Human health effects）

2. 圖例（Examples）

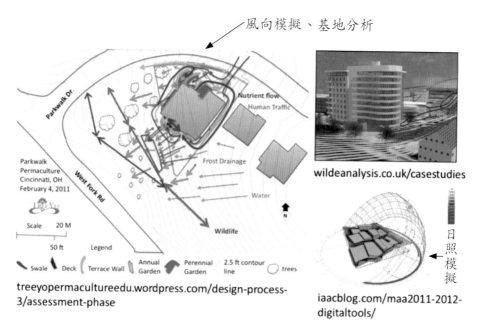

圖 5-2　基地評估之案例

5.2 生物多樣性棲息地（Biodiverse Habitat）

基地開發（Site Development）

• 策略（Strategies）

　　• 保護或恢復自然區域，提供棲息地給當地的動物，人類也可以享
　　　受自然風景與開闊景觀，在曾被破壞的地方種植原生或適應性植
　　　物可以恢復地區和生態系統。（Protect or Restore Sensitive Areas（保
　　　護或復育生態敏感地區）Protect or Restoring natural areas provides
　　　a habitat for local animals and benefits people who will enjoy the views

and open areas. Planting the space of damaged or previously built with native or adaptive plants can restore the area and ecosystems.）

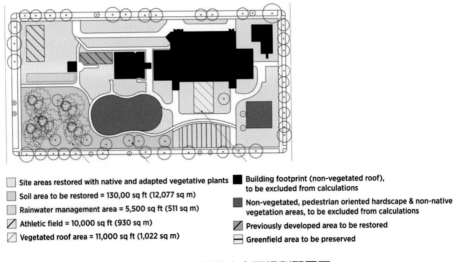

Site areas restored with native and adapted vegetative plants
Soil area to be restored = 130,00 sq ft (12,077 sq m)
Rainwater management area = 5,500 sq ft (511 sq m)
Athletic field = 10,000 sq ft (930 sq m)
Vegetated roof area = 11,000 sq ft (1,022 sq m)

Building footprint (non-vegetated roof), to be excluded from calculations
Non-vegetated, pedestrian oriented hardscape & non-native vegetation areas, to be excluded from calculations
Previously developed area to be restored
Greenfield area to be preserved

圖 5-3　基地開發之全區規劃配置圖

5.3 可及的戶外空間（Accessible Outdoor Space）

策略（Strategies）

1. 縮小建築面積（Build Small）

　　縮小建築面積，增加開放空間。（Minimize the building footprint and maximize open spaces.）

2. 保存開放空間（Preserve Open Space）

　　開放式的空間是一個既有植被與可透水的地方。增加開放空間的數量有助於降低熱島效應，提高雨水管理和保護生態系統。（Open space is an area that is both vegetated and pervious. Increasing the amount of open space helps reduce heat islands, improve rainwater management, and protect ecosystems.）

en.wikipedia.org/wiki/South_Park_Blocks

圖 5-4　人行步道之開放空間

science.howstuffworks.com/environmenta

圖 5-5　屋頂花園層開放空間

提供戶外空間給人實際使用，並以一個或多個項目檢討（Provide outdoor space for physically accessible and be one or more of the following）

- 供行人使用的鋪面或草地，並具備實體元素供戶外社交活動使用（A pedestrian-oriented paving or turf area with physical site elements that accommodate outdoor social activities）
- 供遊憩使用的鋪面或草地，並具備實體元素供戶外體能活動使用（A recreation-oriented paving or turf area with physical site elements that encourage physical activity）
- 擁有生物多樣性的花園空間，可提供一整年不同的風景變化（A garden space with a diversity of vegetation types and species that pro-

vide opportunities for year-round visual interest）

- 作為社區花園或都市農園（A garden space dedicated to community gardens or urban food production）

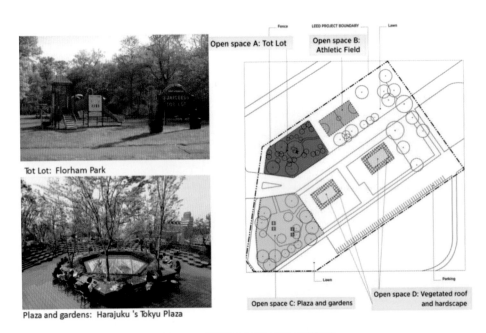

圖 5-6　建築之屋頂花園或綠地

5.4 強化韌性基地設計（Enhanced Resilient Site Design）

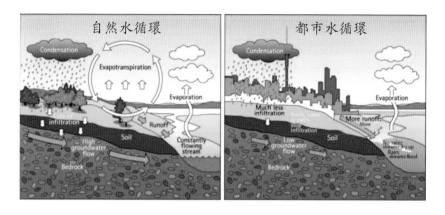

圖片來源：Auckland City Council

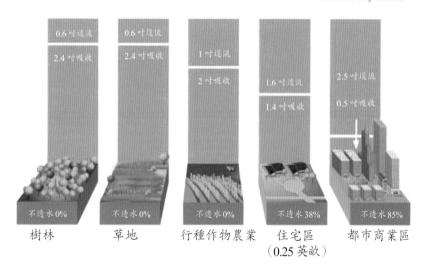

圖 5-7　開發區域與地表逕流的關係

策略（Strategies）

1. 降低暴雨逕流（Reduce Runoff）

　　策略計畫小組考慮建立一個雨水管理計畫用來管理流量。（Consider creating a Rainwater Management Plan that documents which strategies the project team will use.）

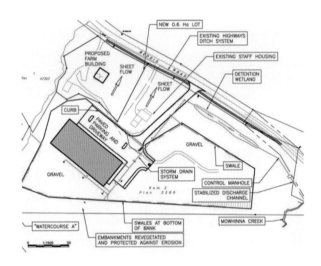

圖 5-8　雨水管理計畫圖

2. 減少不透水鋪面（Reduce Impervious Surfaces）

　　LEED 定義不透水表面為具有小於 50% 的透過性。增加透水表面，如植被屋頂、多孔路面和網格鋪路的面積。（LEED defines an impervious surface as having a perviousness of less than 50%. Increase the area of permeable surfaces, such as vegetated roofs, porous pavement, and grid pavers.）

www.portlandonline.com

圖 5-9　多孔路面材料

3. 減少硬鋪面（Minimize Hardscape）

　　設計硬質鋪面，如停車場、人行道、庭院或替代性表面的傳統鋪面。
（Design hard surfaces such as parking lots, walkways, patios intelligently, or substitute permeable surface for traditional paving.）

www.portlandonline.com　　　　　　www.portlandonline.com

原先：硬鋪面多　　　　　　　後來：草地鋪面

圖 5-10　減少硬鋪面

4. 雨水再利用（Reuse rainwater）

　　收集和儲存雨水，然後重新用來沖洗廁所和灌溉。（Collect and store the rainwater, and reuse it for toilet flushing or irrigation.）

圖 5-11　雨水回收系統、雨水回收桶

5. 低衝擊開發（Low Impact Development）

　　直接徑流進乾水池、雨水花園、植被過濾器、生態草溝保水和潔淨逕流流量的水質。（Direct runoff into dry ponds, rain gardens, vegetated filters , bioswales to hold water and clean the quality of water runoff.）

www.portlandonline.com

圖 5-12　低衝擊開發示意圖

6. 使用以下策略來保持和／或減緩暴雨徑流的速度（Use the following strategies to hold and/or slow the rate of stormwater runoff）

- 透水鋪面（**Porvious** pavement）
- 綠屋頂（Vegetated roofs）

- 生態池（Rain gardens）
- 生態草溝（Bioswales）
- 滯洪池（Retention and detention ponds）

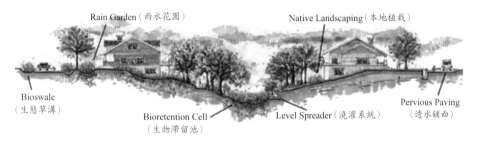

圖 5-13　低衝擊開發之手段

5.5 降低熱島效應（Reduce Heat Island Effect）

熱島效應（Heat Island Effect）

- 從1960年代開始，在世界各大城市發現的地區性氣候現象。（Since the 1960s, a regional climate phenomenon has been observed in major cities worldwide.）
- 由於溫室氣體排放量高，容易吸收熱得。（Due to the high emission of greenhouse gases, these areas tend to absorb more heat.）
- 不論早上或晚上，整體的溫度會較周圍地區來得高。（As a result, both day and night, the overall temperature is higher compared to surrounding rural areas.）

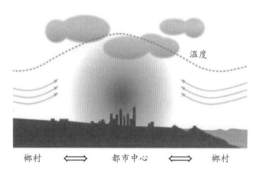

圖 5-14　熱島效應

策略（Strategies）

1. 遮棚／地下停車（Undercover / Underground Parking）

　　停車場設在地下。地下停車場應該有植被、土壤覆蓋層或具有高反射率的覆蓋。（Having parking located underground. Underground parking should have a vegetated or soil cover or a cover that has a high reflectance.）

asbarez.com/81204/　　　greeningthecity.wordpress.com/

圖 5-15　地下停車或屋頂綠化

2. 設置綠屋頂（Green Roofs）

　　設置有植被的屋頂，以減少熱島效應。（Install a vegetated roof to reduce the heat island effect.）

noticias.vidrado.com

圖 5-16　綠屋頂

3. 遮蔭（Shading）

　　景觀設計應納入樹木或者其他植物，可以提供遮蔭。也可以使用太陽能充電板當作停車遮棚的屋頂，不僅可發電，更可以遮陽。（A landscape design should incorporate trees or other vegetation that can provide shade. Solar panels can also create shade for the vehicles underneath.）

alicespringsairport.com.au　　alicespringsairport.com.au

圖 5-17　遮蔭

4. 冷屋頂（Cool Roofs）

　　屋頂使用高反照率（SRI）的涼爽材料，以減少熱島效應的影響。（Specify roof with high albedo（SRI）cool materials to reduce heat island impacts.）

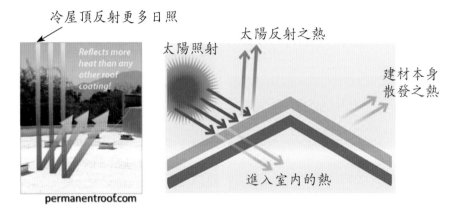

圖 5-18　冷屋頂

5. 冷鋪面（Cool Pavements）

　　黑色瀝青具有低 SRI 係數，而使用高 SRI 係數的材料可以讓鋪面具有更多反射，以減輕太陽熱。（Pavements that are more reflective – those with a higher SRI – can be installed to alleviate the negative properties of lower SRI materials such as black asphalt.）

圖 5-19　冷鋪面

6. 植草磚（Open-grid Pavement）

　　植草磚有開孔，植物會在開孔處生長，對於 LEED 而言，要計算為植草磚的部分，至少需要 50% 以上的面積為草地。（Open grid pavement is a material that has open cells to allow vegetation to grow through it. For LEED, only pavement that is at least 50% pervious is allowed to be counted as open-grid pavement.）

$$植草磚比例：\frac{草地}{總面積} \geq 50\%$$

ochshorndesign.com/cornell　　　　　ochshorndesign.com/cornell

圖 5-20　一般地磚　　　圖 5-21　植草磚

5.6 減少光害（Light Pollution Reduction）

策略（Strategies）

- 照明設計（Lighting design）：良好的照明設施需減少三種形式的光汙染：（Good lighting design involves reducing three forms of light pollution:）
  1. 上照燈（Uplight）
  2. 強光（Glare）
  3. 光侵入（Light trespass）

背光、上照燈、強光（BUG）方法用於 LEED 的光汙染減小。（The

backlight, uplight, glare (BUG) method is used in LEED for light pollution reduction.）

1. 照明控制（Lighting Controls）

　　全自動燈光關閉系統在人們下班後，或人們不需要時自動關閉。這不僅可以減少光汙染，還可以節約能源。（Automatic light shut offs can ensure that lights are not left on after work hours or when they are not needed. This will not only reduce light pollution but also save energy on a project.）

vantage-latam.com/

fasd.hq.nasa.gov/new_office
_lighting.html

圖 5-22　自動光感應偵測　　　　圖 5-23　天光控制開關

2. 照明燈具與設置（Light Fixtures and Placement）

　　使用電腦模擬基地照明，來確保建築物在夜間最低限量的光；使用全遮罩燈具給唯一的光區所需的安全性和舒適。（Design site lighting with computer model to make sure the buildings are lit minimally at night. Use full-cutoff light fixture to only light areas required for safety and comfort.）

圖 5-24　建築外部管理（Building Exterior Management）

3. 景觀維護與蟲害管理（Landscaping and Integrated Pest Management）

有效率管理景觀。制定一個綜合蟲害管理計畫，以改善基地。（Manage landscaping efficiently. Develop an integrated pest management plan to improve sites.）

名詞解釋（Glossary）

- 太陽能反射指數（Solar reflectance index (SRI)/Albedo）

一種衡量材料反射太陽熱量的能力，SRI 越高則這種材料被太陽照射後溫升較小。標準黑色（反射率 0.05，發射率 0.90）為 0，標準白色（反射率 0.80，發射率 0.90）是 100。國際要求 SRI 的計算根據 ASTM 的標準計算方法，換算出材料的 SRI 值。影響材料 SRI 的主要因素包括反射率和發射率。（A measure of a material's ability to reject solar heat, as shown by a small temperature rise. Standard black (reflectance 0.05, emittance 0.90) is 0 and a standard white (reflectance 0.80, emittance 0.90) is 100. For example, a standard black surface has a temperature rise of 90°F (50°C) in full sun, and a standard white surface has a temperature rise of 14.6°F (8.1°C). Once the maximum temperature rise of a given material has been computed, the SRI can be computed by interpolating between the values for white and black.）

反射率和發射率需要在實驗室根據 ASTM 指定的標準測試出來。

一般來說，屋面材料顏色越深，SRI 越低，大部分材料的 SRI 都在 0-100 範圍內。金屬的反射率很高，但是發射率並不高，因此，金屬屋面材料的 SRI 通常只有 40-60 左右，遠遠不及白色塗料的 SRI 高。按照 LEED 的要求，坡屋頂屋面材料的 SRI 不得低於 29，平屋頂不低於 78。屋面的太陽能反射指數越高，那麼夏季升溫就小，從而降低建築製冷能耗，並且緩解城市熱島效應，這種屋面被工程師們稱為涼爽屋面（cool roof）。那些座落在地中海小島上的希臘房屋，大概是利用白色材料反射太陽熱能的最好實例。

下面列舉了一些材料的 SRI：

瀝青——0

灰色水泥瓦——5

白色礦物顆粒瀝青——28

紅色陶土屋瓦——36

鋁——56

白色 EPDM 橡膠——84

白色水泥瓦——90

白色 PVC——100

- 輻射係數（Emissivity）

透過表面純黑體在相同溫度下發出的輻射與物體材料所發出的輻射的比值。（The ratio of the radiation emitted by a surface to the radiation emitted by a black body at the same temperature.）

- 熱發射率（Thermal emittance）

透過標本與由黑體輻射體發出的，在相同的溫度下發射的輻射熱通量的比率。（The ratio of the radiant heat flux emitted by a specimen to that emitted by a blackbody radiator at the same temperature (adapted from Cool Roof Rating Council).）

- 低衝擊開發（Low-impact development (LID)）

一種管理雨水徑流方法，強調用現場的自然特性來保護水質，通過再造的自然土地覆蓋水文流域，並處理徑流量使其接近自然狀態。例如更好的基地設計原則（最大限度地減少土地的干擾，保護植被，減少不透水覆蓋），和設計技術（雨水花園、植被窪地和緩衝、透水鋪面、雨水收集、土壤改良劑）。這些工程的做法，可能需要專門的設計幫助。（An approach to managing rainwater runoff that emphasizes on-site natural features to protect water quality, by replicating the natural land cover hydrologic regime of watersheds, and addressing runoff close to its source. Examples include better site design principles (e.g., minimizing land disturbance, preserving vegetation, minimizing impervious cover), and design practices (e.g., rain gardens, vegetated swales and buffers, permeable pavement, rainwater harvesting, soil amendments). These are engineered practices that may require specialized design assistance.）

- 蒸發散量（Evapotranspiration）

透過植物蒸散的水分與自土壤蒸發的水分總合。（Water lost through transpiration through plants plus water evaporated from the soil.）

- 蒸散率（Evapotranspiration rate (ET)）

單位水蒸散量來自水深單位的植被表面。（The amount of water lost from a vegetated surface in units of water depth.）

- 旱生園藝（Xeriscaping）

一般在嚴重缺水和強烈光照下生長的植物，植株往往變得粗壯矮化。地上氣生部分發育出種種防止過分失水的結構，而地下根系則深入土層，或者形成了儲水的地下器官。另一方面，莖幹上的葉子變小或喪失以後，幼枝或幼莖就替代了葉子的作用，在它們的皮層細胞或其他組織中可具有豐富的葉綠體，進行光合作用。

沙漠地區的很多木本植物，由於長期適應乾旱的結果，多成灌木叢，這在沙漠上生長有很多優越性。

至於許多生長在鹽鹼地的所謂鹽生植物，或旱 - 鹽生植物，由於生理上缺水，也同樣顯出一般旱生的結構。因此旱生園藝成爲一種景觀設計方法，可運用這樣的手法減低植物的需水量。（A landscaping method designed for water conservation so that routine irrigation is not necessary. It includes using drought-adaptable and low-water plants, soil amendments such as compost to conserve moisture, and mulches to reduce evaporation.）

題目（Practice Questions）

1. 在計畫基地上減少不透水表面將會＿＿＿？（Decreasing impervious surfaces on a project site will __?）

   a. 減少滲漏率（Decrease percolation rates）

   b. 減少可飲用水的使用（Reduce potable water usage）

   c. **減少雨水徑流（Reduce stormwater runoff）**

   d. 消除汙水管道（Eliminate sewage piping）

   減少不透水表面，讓透水面增加。透水表面讓水滲透到地下，從而減少雨水徑流。（Decreasing impervious surfaces permits an increase in pervious surfaces. Pervious surfaces allow water to infiltrate the ground, thereby reducing stormwater runoff.）

2. 一個外表具有高太陽能反射指數（SRI）的計畫專案，會對環境有哪些助益？（A project that specifies exterior surfaces with high solar reflectance index (SRI)values is contributing to which environmental benefit?）

  a. 減少熱島效應（Reduced heat island effect）

  b. 支持可再生能源（Support for renewable energy）

  c. 避免夜晚天空行動干擾（Protection of the dark-sky initiative）

  d. 改善雨水性質（Improved stormwater quality）

  太陽能反射指數代表表面反射太陽熱的程度。使用具有較高的太陽能反射指數的材料，會減少基地硬景觀熱量的熱得，從而減少了熱島效應。

  （The solar reflectance index represents how well a surface rejects solar heat. Using materials with high solar reflectance indexes reduces the trapping of heat in site hardscapes, in turn reducing the heat island effect.）

3. 當一個計畫毗鄰於受保護的森林區且該地為許多動植物的棲地，為使照明設備的影響減少，計畫團隊如何安裝外部照明＿＿＿？（A project adjacent to protected forestland that is home to a variety of plant and animal life wants to reduce the impact of its site lighting. To achieve this, the project team installs exterior lighting that__?）

  a. 提供高品味的外觀裝飾。（Provides for tasteful, decorative appearance.）

  b. 降低對夜間的安全性。（Reduces the need for night-time security.）

  c. 充分照亮夜空。（Adequately illuminates the night sky.）

  d. 不任意侵入到相鄰的領地。（Does not trespass onto adjacent properties.）

  光害侵擾是指不必要的光溢出到相鄰的領地。為了最大限度地減少其對夜間周圍的影響，光侵入必須控制。（Light trespass is the unwanted spill age of Light onto adjacent properties. To minimize its impact on nocturnal environs, light trespass must be controlled.）

4. 一個團隊要提供計畫基地優質的開放空間。應考慮以下那個策略？（A team wants to provide quality open space on the project site. Which of the following strategies should it consider?）

a. 毗鄰計畫邊界的一個社區花園（A community garden adjacent to the project boundary）

b. 所有物周圍的步行道（A walking trail around the property）

c. 公司獨家經營私人花園（A private garden for company exclusive）

d. 滯洪池（A water-detention pond）

步行道在計畫的基地上鼓勵社交和體育活動，所以可算作開放空間。（A walking trail on the project site encourages social interaction and physical activity. So it would count as open space.）

5. 下列哪一項會提高市區的熱島效應？（Which of the following can increase the heat island effect in urban areas?）

a. 大面積的草坪（Large areas of turf grass）

b. 減少高樓和窄巷的空氣流通（Reduced air flow from tall buildings and narrow streets）

c. 地下停車場（Underground parking）

d. 來自樹的硬景觀遮蔽（Hardscape shaded by tree）

熱島效應的主要原因是深色表面，如屋頂或暗的瀝青路面，會吸收熱量並散發入周邊地區。（The primary cause of the heat island effect is dark surfaces such as rooftops or dark asphalt pavement that absorb heat and radiate it into the surrounding areas.）建築和狹窄的街道之間的空氣流通減少也會增加熱島效應。其他造成熱島效應的原因包括冷氣、汽車的排氣

及晴朗的天氣。（Reduced air flow between buildings and narrow streets also increase the effect. Other causes of the heat island effect include air-conditioners, vehicle exhaust, and calm and sunny weather.）

6. 以下哪位計畫團隊成員通常是負責建築活動汙染防治的開發侵蝕和沉積控制（ESC）規劃？（Which of the following project team member is typically responsible for developing the erosion and sedimentation control (ESC) plan for construction activity pollution prevention?）

   a. 土木工程師（**The civil engineers**）

   b. 業主（The project owner）

   c. LEED 綠建築專業認證人員（The LEED AP）

   d. 建築師（The architect）

   土木工程師通常負責該計畫。景觀設計師或總承包商可能只負責計畫中的一小部分。（The civil engineer is usually responsible for the plan. The landscape architect or general contractor may work on the plan's development.）

7. 採用開放式網格路面可以幫助計畫達成哪種 LEED 領域？（The use of open grid pavement can help a project achieve points in what LEED areas?）

   a. 戶外減少用水量（Outdoor Water Use Reduction）

   b. 減少熱島效應（**Heat Island Reduction**）

   c. 開放空間（Open Space）

   d. 雨水管理（**Rainwater Management**）

   開放式網格路面是路面面積小於 50% 的不透水鋪面，並在開放的單元格包含植被（通常為草）。在開放單元格的植被取代吸熱鋪面，可以降低吸收的熱量。（Open grid pavement is pavement that is less than 50% impervious and contains vegetation in the open cells. The vegetation in the open

cells **replaces heat absorbing surfaces** just like any other plant.）
開放式網格路面透過減少不透水表面的數量管理雨水徑流。（Open grid pavement helps **manage runoff** by reducing the quantity of impervious surfaces.）

8. 什麼樣的設計策略將促進生物最多樣性？（What design strategy would promote biodiversity the most?）

a. 建造一個乾池塘並種植外來入侵植物（Installing a dry pond planted with invasive plants）

b. 增加開放空間並用草坪覆蓋它（Increasing open space and covering it with turf grass）

c. **種植了各種本土植物（Planting a variety of indigenous plants）**

d. 造一個岩石花園（Installing a rock garden）

天然的（或原生、本地的）植物是那些在一個地區自然生長或已在一個地區多年。原生植物需要更少的水、肥料和病蟲害防治。這些植物可以是喬木、灌木、花草或草。自適應植物，在當地的氣候表現良好的非原生植物。本地和適應性植物需要更少的水、更抗病，因爲它們適合於該地區的通常降雨量、土壤和溫度。（Native (or indigenous) plants are those that grow naturally in an area, or that have been in an area for many years. Native plants require less water, fertilizer, and pest control. These plants can be trees, shrubs, flowers, or grasses. Adaptive plants are non-native plants that perform well in the local climate. Native and adaptive plants require less water, and are more disease resistant because they are suited to the region's usual rainfall, soil, and temperature.）

一種植物如果在一個地區經過長時間的自然生長和演化，適應當地的地質、水文和氣候，我們就稱之爲當地的原生植物（native plant）或本土

植物。它們和當地的其他動植物長期共同演化，成爲許多野生物的棲息地。

種植原生植物的好處：

1. 原生植物已經適應生長地的地質、水文和氣候，因此可以減少肥料、灌水以及農藥的使用，維護工作比較簡單，而且成本低廉。

2. 原生植物和生長地的其他生物長期共同演化，因此爲許多野生物提供食物和棲息所。

3. 原生植物可以增加當地的生物多樣性。

4. 原生植物會引來其他原生生物遷入，提供更豐富的欣賞和觀察經驗。

5. 原生植物多具深根性，比較可以抵擋風害和旱害，可增加土壤貯水及水土保持功能。

9. 下列哪個現象會因爲雨水逕流增加而導致？（Which of the following can happen as a result of increased rainwater runoff?）

a. 提高可飲用水的使用（Increased potable water use）

b. 優養化（Eutrophication）

c. 增加熱島效應（Increased heat island）

d. 減少能源表現（Decreased energy performance）

優養化（Eutrophication）河流或湖泊的植物養分過量時所產生的汙染過程，會導致藻類和其他水生植物過量生長（is the ecosystem response to the addition of artificial or natural substances, such as nitrates and phosphates from fertilizers or sewage, to an aquatic system.）

額外的硬景觀會促使水質優養化和危害水生生態系統與物種。（Additional hardscapes can contribute to eutrophication and harm aquatic ecosystems and species.）

第六章　用水效率（Water Efficiency）

學習目標（Learning Objectives）

- 室外與澆灌用水（Outdoor Water Use Reduction）
- 室內用水（Indoor Water Use Reduction）
- 建築裝設水表（Building-Level Water Metering）
- 冷卻水塔用水（Cooling Tower Water Use）

影響（Impacts）

- 遠離自然水體。（Withdrawals from natural water bodies.）
- 人造環境阻礙水體自然迴流。（Built environment makes it increasingly difficult for water to naturally recharge the system.）
- 水供應減少，用水費用提高。（Water supply dwindles; water not underpriced anymore.）
- 減少用水量，減少用能費用。（Less water used, less energy consumption）
- 水是人類存在的基礎。（Water is fundamental to our existence.）
- 為後代子孫保護安全乾淨的水資源。（Ensure safe, clean water for future generation.）

圖 6-1　Triple Bottom Line〔三重底線（三大原則）〕

用水效率得分的目標（Goals of the Water Efficiency Credits）

- 減少建築和景觀所需的用水數量（Reduce the quantity of water needed for a buildings and landscaping）
- 減少自來水用量（Reduce municipal water use）
- 減少廢水處理需要（Reduce the need for treatment of waste water）

優先事項（Priority）

- 對於能源利用效率和用水效率，LEED 需要一個高效率的做法。然後，再尋找其他方式來減少使用。因此，開源節流是必要的手段。（For both energy efficiency and water efficiency LEED requires an efficiency first approach. After efficiency, then look for other ways to reduce use.）

用水效率可由下列幾個面向來討論：

- 室外用水（Outdoor water）
- 室內用水（Indoor water）
- 作業用水（Process water）

圖 6-2　各式用水器具

6.1 水表量測與報告（Water Metering and Reporting）

策略（Strategies）

1. 選擇適當植栽（Appropriate Plant Selection）

原生和適應性植物需要較少的水，而且更有抗病蟲能力，因為它們已適應本地區的慣常降雨、土壤和溫度。類似的植物應進行分類，以便最大限度地提高用水效率。（Native and adaptive plants require less water and are more disease resistant because they are suited to the region's usual rainfall, soil and temperature. Similar plants should be grouped together to maximize water efficiency.）

wcasohio.org

圖 6-3　本地植栽手冊

　　避免使用外來入侵植物，那些外來植物生長迅速、侵略性強，並排擠其他植物。多年生的花卉優於一年生植物，因為多年生植物每年都會生長，且較不需要澆水。（Avoid the use of invasive plants, which are plants that grow quickly and aggressively, spreading and displacing other plants. Perennial flowers are preferred to annuals, because perennials will come back year after year and will require less watering.）

hickoryhollowlandscapers.com

圖 6-4　多樣化植栽

2. 旱生園藝（Xeriscaping）

對減少灌溉是非常有效的，通常是指土地可以恢復之前的狀態。（Eliminating irrigation is incredibly efficient and usually means land is taken back to a historic content.）

xeriscaped.com

distinctbuild.ca

圖 6-5　旱生植栽

3. 護根（Mulching）

在天氣溫暖時安裝 1～3 英寸的覆蓋物保持植物根系涼爽濕潤，有助於防止水分蒸發。（Installing 1-3 inches of mulch keeps a plants root system cool and moist in warmer weather, helping prevent evaporation.）

圖 6-6　護根

4. 減少草地面積與單一栽培（Reduce Turf Grasses and Monocultures）

　　草坪的草需要大量的水。草坪的草通常是單一作物。生物多樣性不鼓勵使用單一種植。（Turf grasses require large amount of water. Turf grass is an example of a monoculture, or a single species of plant. Monocultures do not promote biodiversity.）

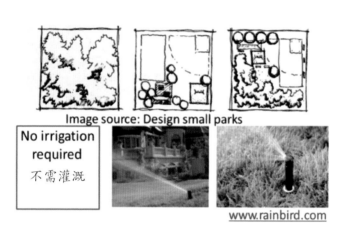

圖 6-7　　草坪規劃與灌溉設備

5. 微滴灌系統（Drip Irrigation）

　　常規灌溉具有約 65% 的灌溉效率。滴灌具有 90% 的灌溉效率。滴灌是灌溉用水最有效的形式，因爲沒有地表徑流水散失。（Conventional irrigation has an irrigation efficiency of about 65%. Drip irrigation has an irrigation efficiency of 90%. Drip irrigation is the most water efficient form of irrigation because there is no surface runoff water.）

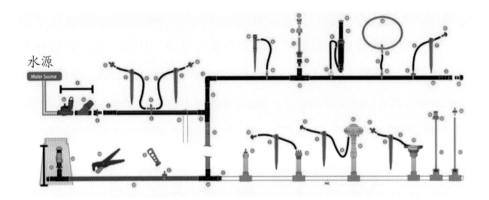

圖 6-8　一般各式灌溉工具其可能漏水之位置

6. 澆灌時程（Scheduling）

　　深度澆灌使得植物的根能夠向下紮根，更深入土壤。其中，可以發現隨著時間的推移更加濕潤。（Deep watering forces the plant's roots to push further down into the soil where more moisture can be found over time.）

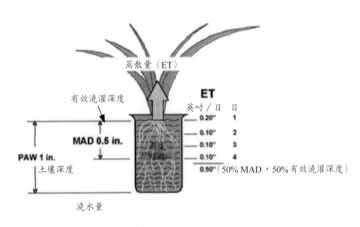

圖 6-9　土壤澆灌示意圖

7. 依天候調整的控制器（Weather Based Irrigation Controllers）

　　天氣或者根據感應器的灌溉控制技術採用當地的氣候和景觀條件，以歷史的天氣數據的情況調整實際灌溉時程及頻率。（Weather- or sensor-based irrigation control technology uses local weather and landscape conditions to tailor irrigation schedules to actual conditions on the site or historical weather data.）

感應器

圖 6-10　根據濕度自動灑水器

　　代替根據預先設定的時間表澆灌，先進的灌溉控制器允許配合植物對水的需求灌溉。（Instead of irrigating according to a preset schedule, advanced irrigation controllers allow irrigation to match the water requirements of plants.）

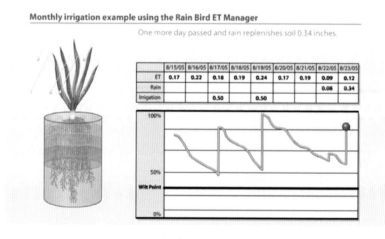

圖 6-11　自動化澆灌系統之實例

8. 景觀養護（Landscape Maintenance）

　　照顧和日常維護的草坪建立一維護整理計畫，維護策略包括提高割草機效率、不要太常割草、除草以維護供水系統。（Establishes procedures for cutting, caring for and routine maintenance of lawn and landscaping. Maintenance strategies include Raise Mowers, Leave the Clippings, and Maintain the Watering System.）

圖 6-12　　景觀維護

9. 使用雨水或灰水（Use Rainwater and/or Graywater）

　　雨水可以收集在蓄水池、桶或儲水箱。這種方法被稱為雨水收集。（Rainwater can be collected in cisterns, barrels, or storage tanks. This approach is called rainwater harvesting.）

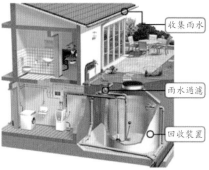

收集雨水

雨水過濾

回收裝置

自來水
（Potable Water）

灰水（**Greywater**）
從浴缸、淋浴、浴室水槽、洗衣
機和洗衣桶中排出的廢水。
（Waster water discharged from
bathtubs, showers, bathroom sinks,
washing machines and laundry
tubs.）

灰水，中水
（Greywater）

圖 6-13　灰水之流程

　　如果灰水被正確過濾，重複使用它進行灌溉，從健康的角度來看是安
全且有效的利用。（If gray water（又名 grey water）is filtered properly, reusing it for irrigation and further conveyance is safe from a health perspective.）

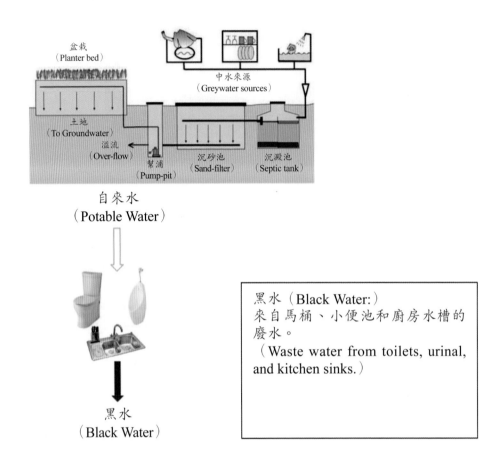

圖 6-14　黑水之流程

6.2 最低用水效率（Minimum Water Efficiency）

1. 使用高效節水用具（Use Efficient Fixtures）

　　（雙沖水 / 高效率廁所、無水小便器、堆肥廁所、低流量蓮蓬頭和水龍頭、水龍頭低流量充氣器和 / 或動作感知器）（Dual-flush/High-Efficiency Toilets, Waterless Urinals, Composting Toilets, Low-flow Showerheads and Faucets, Faucets with Low-flow Aerators and/or Motion Sensors）

噴霧式龍頭	堆肥系統	兩段式沖水馬桶
（onthejobbob.com）	（tlc.howstuffworks.com）	（homedosh.com）

圖 6-15

　　WaterSense 是由美國環境保護署（EPA）認可的方案，幫助消費者識別高性能節水馬桶，可以減少水和廢水利用、降低成本和節省國家的水資源，類似於台灣的省水標章。（WaterSense is a program sponsored by the United States Environmental Protection Agency (EPA) that helps consumers identify high-performance, water-efficient toilets that can reduce water and wastewater use, costs, and conserve the nation's water resources.）

用水器具之用水基準值

器具或配件	基準（IP 單位）	基準（SI 單位）
馬桶 *	1.6 gpt	6 lpf
小便斗 *	1.0 gpt	3.8 lpf
公用洗手台（盥洗室）水龍頭	60 psi** 壓力下 0.5gpm，除私人使用的用水器具之外的其他所有用水器具	415 kPa 壓力下 1.9 lpm，除私人使用的用水器具之外的其他所有用水器具
私人使用浴室水龍頭	60 psi 壓力下 2.2 gpm	415 kPa 壓力下 8.3 lpm
廚房水龍頭（專用於灌裝操作的水龍頭除外）	60 psi 壓力下 2.2 gpm	415 kPa 壓力下 8.3 lpm

器具或配件	基準（IP 單位）	基準（SI 單位）
淋浴噴頭 *	每個淋浴間 80 psi 壓和下 2.5 gpm	每個淋浴間 550 kPa 壓力下 9.5 lpm

* 該產品類型可使用 Waler Sense 標籤
　gpf ＝ 每次沖水加侖數
　gpm ＝ 加侖每分鐘
　psi ＝ 磅每平方英寸

lpf ＝ 每次沖水升數
lpm ＝ 升分鐘
kPa ＝ 千帕

2. 安裝水錶（Install Water Meters）

在不同位置安裝水錶，將有助於測量和驗證廢水的使用。（Installing water meters for different areas will help in measurement and verification of waste usage.）

ctm.co.uk　　　　　　　indiawaterportal.org

圖 6-16　安裝水錶

3. 使用雨水或灰水（Use Rainwater and/or Graywater）

對於居家使用中水可減少 30% 用水量，在商業規模的建築上有更大的節水潛力。（Recycled graywater could provide an estimated 30% reduction in water use for household, and greater on a commercial scale.）

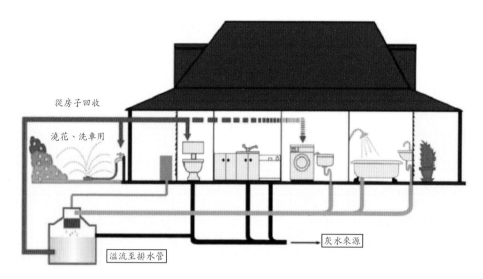

從房子回收

澆花、洗車用

灰水來源

溢流至排水管

圖 6-17　居家用水示意圖

使用作業用水（Use process water）

自來水
（Potable Water）

作業用水
（Process Water）

作業用水（Process Water）：
用於工業過程和建築系統，它包括：（used for industrial processes and building systems,it includes：）
- 機電系統的補水（Make up water for mechanical system (Cooling towers, boilers, and chillers)）
- 實驗室或醫院的系統用水（Systems that use water in laboratories or hospitals）
- 商業廚房的器具：蒸飯、製冰、洗碗機等（Commercial cooking applications – food steamers, ice machines, dishwashers, etc.）

圖 6-18　使用作業用水

處理過的水代替非飲用水是減少水的使用的方法之一。收集水或回收的水可以用於滿足加工用水的需要。（Substituting non-potable water for processed water is one way to reduce water use. Captured water or reclaimed water can be used to meet the needs for process water.）

冷卻塔：化學控制冷卻塔的管理必須用水有效率且安全。蒸發冷卻塔可以比一個標準冷卻塔節約高達 20%，這取決於氣候和配置。有效的監測冷卻塔可以創建一個更有效的系統。（Cooling tower- Chemically controlled cooling tower management must be water efficient and safe. An evaporative cooling tower can conserve up to 20% of the water use of a standard cooling tower, depending on climate and configuration. Effective monitoring of cooling towers can create a more efficient system.）

6.3 水表量測與漏水檢測（Water Metering and Leak Detection）

- 比建築物的用水基準計算值少用 20% 的水。沖洗和流量是根據能源政策法案（EPAct）1992 年的標準。（Use 20% less water than the water use baseline calculated for the building. The flush and flow rates are based on the EPAct 1992 standard.）

- 基準用水量是基於估計居住者使用（全時間計算）與水設備和家具。（The baseline water usage is based on estimated occupant usage (Full Time Equivalent calculations) and the water fixtures and fittings.）

能源政策法案（Energy Policy Act）

1992 年的能源政策法案（EPAct）建立節約用水標準，廁所、淋浴噴頭、水龍頭等。（The Energy Policy Act of 1992 (EPAct) established water conservation standards for water closets, shower heads, faucets, and other.）

商業住宅空間	目前基準
商業，馬桶	1.6 gallons per flush (gps)* Except blow-out fixtures: 3.5 (gpf)
小便斗	1.0 (gpf)
水龍頭	2.2 gpm 60 psi 私人區 0.5 gpm 60 psi 公共區
沖洗閥	沖洗率 1.6 gpm

住宅	目前基準
馬桶	1.6 加崙 / 沖一次
水龍頭	2.2 加崙 / 沖一次 60 psi 的壓力下
廚房水龍頭	
蓮蓬頭	2.5 加崙 / 分鐘 80 psi 的壓力下

gpf = gallon per flush（沖一次多少加崙）
gpm = gallon per minute（加崙 / 分鐘）

6.4 加強用水效率（Enhanced Water Efficiency）

整體專案用水量（Whole Project Water Use）

為追求此途徑，專案團隊必須建立用水基線並創建擬定的用水模型。得分根據相對於基準線的減少量獲得。（To pursue this pathway, project teams must develop a water use baseline and create a proposed use model. Points are achieved based on reductions from the baseline.）

裝置和配件──計算減少量（Fixture and Fittings-Calculated Reduction）

進一步減少裝置和配件的用水量，基於 WEp：最低水效益的計算基線、最低裝置和配件效能，效能路徑─計算減少量。使用替代水源可

在達到前提條件水平之上獲得額外的飲用水節約。（Further reduce fixture and fitting water use from the calculated baseline in WEp: Minimum Water Efficiency, Minimum Fixture and Fittings Efficiency, Path 2. Performance Path – Calculated Reduction. Additional potable water savings can be earned above the prerequisite level using alternative water sources.）

- 計算冷卻塔最大循環，將每個參數的允許最大濃度水平除以飲用補充水分析中每個參數的實際濃度水平。限制冷卻塔的循環次數，以避免超過任何這些參數的最大值。與冷卻塔水接觸的水系統構建材料應能在此循環數範圍內運行和維護。（Calculate the maximum number of cooling tower cycles by dividing the maximum allowed concentration level of each parameter by the actual concentration level of each parameter found in the potable makeup water analysis. Limit cooling tower cycles to avoid exceeding maximum values for any of these parameters. The materials of construction for the water system that come in contact with the cooling tower water shall be of the type that can operate and be maintained within the cycles.）

優化冷卻用水量（Optimize Water Use for Cooling）

要符合此的資格，根據 ASHRAE 90.1-2019 或 90.1-2022 附錄 G 表 G3.1.1-3，為建築指定的基線系統必須包含冷卻塔。在水冷卻機組系統上實現超越具有軸向變速風扇冷卻塔和最大漂移 0.002% 的三個冷卻塔循環的水效能。（To be eligible for Option 2, the baseline system designated for the building using ASHRAE 90.1-2019 or 90.1-2022 Appendix G Table G3.1.1-3 must include a cooling tower (systems 7, 8, 11, 12, and 13). Achieve increasing levels of cooling tower water efficiency beyond a water-cooled chiller system

with axial variable-speed fan cooling towers having a maximum drift of 0.002%
of recirculated water volume and three cooling tower cycles.）

用水再利用（Water Reuse）

用水回收準備系統（Path 1. Reuse-Ready System）

安裝供水系統，以便爲以下一個或多個最終用途提供回收水或替代
水。應爲與最終用途相關的處理設備預留空間。（Install a water supply sys-
tem to allow the supply of reclaimed or alternative water to one or more of the
following end-uses. Space shall be provided for treatment equipment as applicable to end uses.）

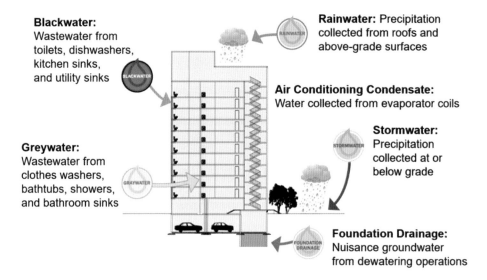

Blackwater:
Wastewater from
toilets, dishwashers,
kitchen sinks,
and utility sinks

Rainwater: Precipitation
collected from roofs and
above-grade surfaces

Air Conditioning Condensate:
Water collected from evaporator coils

Stormwater:
Precipitation
collected at or
below grade

Greywater:
Wastewater from
clothes washers,
bathtubs, showers,
and bathroom sinks

Foundation Drainage:
Nuisance groundwater
from dewatering operations

圖 6-19　用水再利用系統（Potable Water Reuse Systems）

來源：San Francisco Public Utilities Commission.

替代水源（Alternative Water Sources）

爲建築和基地的一個或多個最終用途整合以下室內、室外和／或作業用水的水重用策略。（Incorporate one of the following water reuse strategies for indoor, outdoor, and/or process water that meets the needs of one or more end-uses for the building and grounds.）

基地內水再生系統（Onsite water reuse system）

自來水供應回收水（Municipally supplied reclaimed water）

合格最終用途包括灌溉、沖洗設備、冷卻塔或鍋爐等補充水系統，或其他作業用水需求。（Eligible end-uses include irrigation, flush fixtures, makeup water systems, such as cooling towers or boilers, or other process water demand.）

名詞小教室

全時當量（FTEs）

在減少用水量的部分，以及一些其他的 LEED 評分需要計算的項目，必須建立基準，計算人數的方法稱爲全時當量（FTE）。（The water use reduction prerequisite as well as some other LEED credits use a variable called Full-Time Equivalent (FTE) for calculating baselines and design case values.）

全時當量爲計算在建築物的居民人數，全日制及部分時間的上班族人數，以及短暫居住者的人物。（FTE is a way of calculating the number of residents in a building, the number of full time and part-time office workers, and the number of transient occupants.）

建築物的居住者標識的類型：（Building occupants are identified by type:）

- 全職員工（Full-time staff）
- 兼職員工（Part-time staff）
- 短暫的居住者，如學生、客戶、訪客（Transient occupants (students, customers, visitors)）
- 居民（Residents）

名詞解釋（Glossary）

- 雨水回收（Harvested rainwater）

  沉澱與收集雨水，用於建築物室內的使用或是室外的灌溉，或兩者兼而有之。（Precipitation captured and used for indoor needs, irrigation, or both.）

- 可飲用水（Potable water）

  符合或超過美國環保署的飲用水水質標準，由具有管轄權的國家或地方政府提供的人類生活用水；它可能會透過水井或市政供水（自來水）系統提供。（Water that meets or exceeds EPA's drinking water quality standards and is approved for human consumption by the state or local authorities having jurisdiction; it may be supplied from wells or municipal water systems.）

- 堆肥式馬桶（Composting toilet systems）

  透過細菌或微生物處理人類排放的排泄物轉化成有機成分。（Dry plumbing fixtures that contain and treat human waste via microbiological processes.）

- 微滴灌澆灌（Drip irrigation）

  一種高效率的方法，水在低壓下透過掩埋的水管輸送。在網狀的水管中，水從孔洞流到土壤中。滴灌是一種微量澆灌，是較省水的灌溉方式。（A high-efficiency method in which water is delivered at low pressure through buried mains and submains. From the submains, water is distributed to the soil from a network of perforated tubes or emitters. Drip irrigation is a type of microirrigation.）

- 等值全職員工（Full-time equivalent (FTE)）

  代表一個全職一週工時 40 小時的人力。兼職或加班的人力會根據代表了每週 40 個小時為一個 FTE 的方式計算，常用於建築物中的用水量估算。（A regular building occupant who spends 40 hours per week in the project building. Part-time or overtime occupants have FTE values based on their hours per week divided by 40. Multiple shifts are included or excluded depending on the intent and requirements of the credit.）

- 回收廢水（Reclaimed water）

  經過處理和再利用清潔後的廢水。（Wastewater that has been treated and purified for reuse.）

題目（Practice Questions）

1. 什麼是室內用水基準線的首要標準？（What is the primary standard used to establish the baseline case for indoor water use?）

   a. 清潔水法（The Clean Water Act）

   b. 1992 年能源政策法案（Energy Policy Act of 1992）

   c. ASHRAE 標準 90.1（ASHRAE Standard 90.1）

   d. 國家環境政策法案（National Environmental Policy Act）

許多用水裝置遵循的基本用水的能源政策法案（EPACt 1992）的規定。其他用水裝置遵循如通用管道設備編碼（UPC）的標準。（Many water-using fixtures follow the Energy Policy Act (EPAct) guidelines for baseline water use. Other water-using fixtures follow standards such as the Universal Plumbing Code (UPC).）

2. 冷卻塔使用再生水減少了什麼用水的使用？（What is reduced when a project uses reclaimed water in its cooling towers?）

   a. 飲用水使用（**Potable water use**）

   b. 過程用水（Process water use）

   c. 室內水管用水（Indoor plumbing water use）

   d. 非飲用水使用（Non-potable water use）

   e. 灌溉用水（Irrigation water use）

   使用再生水的過程，可以減少專案對飲用水需求。（Projects that use reclaimed water for process uses reduce their demand on municipally supplied potable water.）

3. 如何可以減少或消除飲用水及灌溉用水用量？（選擇兩個）（How can potable water use for irrigation be reduced or eliminated (select two)?）

   a. 安裝分水錶（**Install sub meters**）

   b. 選擇適合當地的植物（**Select locally adapted plants**）

   c. 提高草坪的覆蓋範圍（Increase the coverage of turf grass）

   d. 使用有機肥料（Use organic fertilizers）

   e. 選擇非外來種植物（Select noninvasive plants）

   分水錶和適應當地的植物有助於減少用水量。分水錶可以監測水管在何處洩漏或消耗較大的部分，因而從其改善。採用適合當地種植面積來種植適當植物，因為這些植物都適合當地氣候，一旦種植可以用很少甚至

無水來進行灌溉。（Sub meters and locally adapted plantings both contribute to water use reduction. Sub meters ensure that water use can be tracked and leaks or overwatering quickly mitigated. The use of locally adapted plantings allows further reduction, because these plants are suited to the local climate and, once established, can be sustained with little or no ongoing irrigation.）

4. 非飲用的水通常適用於下列哪些用途（選擇兩個）？（Nonpotable water is typically suitable for which of the following uses (select two)?）

   a. 製冰（Ice making）

   b. 飲用水（Drinking）

   c. 淋浴（Showers）

   d. 廁所沖水（Toilet flushing）

   e. 灌溉（Irrigation）

   非飲用水根據定義，不適合食用，所以經常用於植物澆水和廢料運輸，但是不可用於飲用、製冰、洗澡，或與人體直接接觸。（Nonpotable water, by definition, not suitable for consumption, so although it is often acceptable to use for plant watering and waste transport, it is not usable for drinking, ice making, or bathing.）

5. 市政供應再生水被認為是？（Municipally supplied reclaimed water is considered?）

   a. 免費水（Free water）

   b. 非飲用水（Nonpotable water）

   c. 作業用水（Process water）

   d. 黑水（Blackwater）

   許多政府機關提供飲用水（處理供人類消費）和非飲用水（基本處理，但不跟飲用水相同標準）。這種非飲用水經常來自於雨水回收。（Many municipalities supply both potable water (treated for human consumption) and nonpotable water (typically treated, but not to the same standard as potable water). This nonpotable water is often reclaimed from sources such as storm water.）

6. 作業用水的使用可透過＿＿＿減少？（選擇兩個）（Process water use can be reduced by __?（select two））

   a. 安裝水錶（**Installing sub meters**）

   b. 安裝低流量蓮蓬頭（Installing low-flow showerheads）

   c. 使用高效率的灌溉技術（Using high-efficiency irrigation technologies）

   d. 使用能源之星認證的洗衣機（**Using ENERGY STAR-certified clothes washers**）

   作業用水用於工業系統例如 HVAC ，以及對某些業務的操作，例如衣服的洗滌和洗餐具。分水錶可以追蹤消耗，並允許早期識別低效率或洩漏。能源之星認證的洗衣機具有低水耗的功能，從而確保他們有效地使用水。（Process water is used for industrial systems such as HVAC, as well as for certain business operations, such as clothes washing and dish washing. Submetering can track consumption and allow for the early identification of inefficiencies or leaks. ENERGY STAR-certified clothes washers have low water factors, thereby ensuring that they use water efficiently.）

7. 廁所水龍頭的計量單位是下列何者？（What unit of measurement is used for a lavatory faucet?）

   a. 設計效率（Design efficiency）

   b. 每分鐘加侖／升（**Gallons/liters per minute**）

   c. 加工用水率（Process water rate）

   d. 每次沖水加侖／升（Gallons/liters per flush）

每分鐘加侖是量測器具流量的單位，如水龍頭。從 1992 年的能源政策法的標準，基準流速洗手台龍頭 2.2 加侖（每分鐘加侖）。（Gallons per minute is a measurement of flow fixture such as faucets. From the EPAct standard of 1992, the baseline flow rate for lavatory faucet is 2.2 gpm (gallons per minute).）

8. 爲了減少室內用水，產品應該選擇？（選擇兩個）（What types of products should be selected for reducing indoor water use?（Choose 2））

   a. WaterSense

   b. Green-e

   c. 能源之星（ENERGY STAR）

   d. ISO

   WaterSense 是一種 EPA 認證，頒發給相較於其他裝置，使用較少水的設備。（WaterSense is an EPA certification awarded to fixtures that use less water than comparable fixtures.）

   合格於 ENERGY STAR 的洗衣機，使用少於一般洗衣機約 37% 的能源及節省超過 50% 以上的用水量（ENERGY STAR qualified clothes washer use about 37% less energy and use over 50% less water than regular washers）

9. 居住人員數量和計算可能需要以下哪種文件？（選擇兩個）（Occupancy calculations may be needed for which of the following project documentation?（choose 2））

   a. 室內使用水（Indoor water use）

   b. 用水計量（Water metering）

   c. 周圍密度（Surrounding density）

   d. 自行車設施（Bicycle facilities）

   室內使用水可能需要居住者數量計算，以確定基準線和設計的情況下使

用。（Indoor water use may require occupancy calculations to determine the baseline and design case usage.）

當計算住戶和遊客的自行車停放數量所需的占用空間是必需的（When calculating the number of bicycle storage units required for occupants and visitors the occupancy count is needed）

10.哪一項是灰水的來源？（Which of the following are sources of graywater?（choose 2)）

a. 洗衣機（**Washing machines**）

b. 廁所（Toilets）

c. 水井（Wells）

d. 淋浴（**Showers**）

根據當地法規洗衣機通常被歸類為灰水。（Washing Machines is usually classified as graywater depending on local code.）

取決於當地的規範，淋浴水往下排水被分類為灰水。（Shower water that goes down the drain is then classified as graywater, depending on local code.）

第七章　能源與大氣
（Energy and Atmosphere）

學習目標（Learning Objectives）

- 基礎進階功能驗證（Fundamental & Enhance commissioning）
- 能源模擬（Optimize energy performance）
- 能源量測、安裝分電表（Build-level & Advanced energy metering）
- 再生能源使用（Renewable energy production）
- 綠色電力（Green Power）

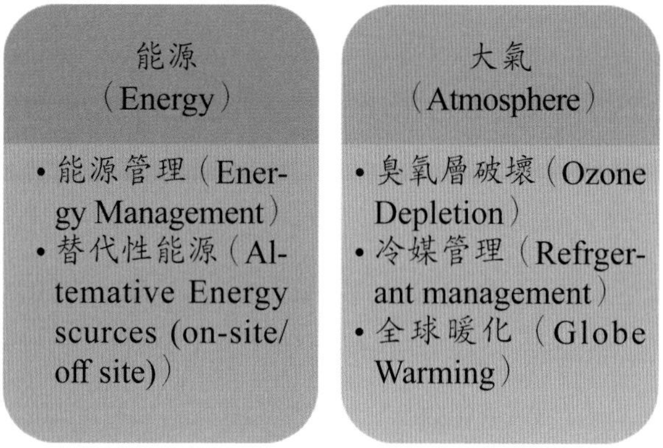

能源 （Energy）	大氣 （Atmosphere）
• 能源管理（Energy Management） • 替代性能源（Altemative Energy scurces (on-site/off site)）	• 臭氧層破壞（Ozone Depletion） • 冷媒管理（Refrgerant management） • 全球暖化（Globe Warming）

圖 7-1

能源管理工具（Energy Management Tools）

1. 功能驗證（調試）（Commissioning (CX)）

　　系統運行的過程中確保建築執行狀態是按照設計意圖和業主的需求執行。在建築物設計施工時，輔以完整嚴謹的功能驗證是防止設計及安裝出

問題的重要手段。功能驗證可由一個團隊組成，接受業主委託。（Systematic process of assuring that a building performs in accordance with the design intent and the owner's operational needs.）

2. 能源之星數據管理器（Energy Star Portfolio Manager）

以整個區域的建築物來說，一個互動性的能源管理工具是用來追蹤和評估能耗和水耗。（An interactive energy management tool for tracking and assessing energy and water consumption across an entire portfolio of buildings.）

3. 電腦模擬軟體（DOE's EnergyPlus Computer Program DOE's）

在早期設計階段使用軟體執行建築節能與分析設計，推估其隨著時間推移後建築物在各階段的能源表現。（A Software take inputs and determine the energy behavior of a building's systems over time, performs building energy analysis and design at early design stage.）

4. 美國冷凍空調協會 ASHRAE 90.1-2010 標準（ASHRAE Standard 90.1）

本標準在建築物能耗設計上設置最低的需求。（The standard that sets minimum requirements for the energy efficient design of buildings.）

7.1 營運碳排放預測與脫碳計畫（Operational Carbon Projection and Decarbonization Plan）

策略（Strategies）

使建築利害相關人能夠預測其當前設計決策對專案長期運營碳排放的影響，並確保利害相關人從專案開始就計畫實現低碳結果。（To enable building stakeholders to visualize how their current design decisions will impact their project's long-term operational carbon emissions and to ensure that stakeholders are planning for low carbon outcomes from the project's inception.）

設計分析（Design Analysis）

- 現場能源預測（Site Energy Prediction）
- 碳排放預測審查（Review Carbon Projection）
- 脫碳計畫（Decarbonization Plan）

路徑 1：電氣化設計（Path 1. Design for Electrification）

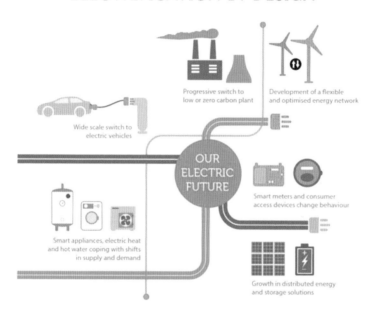

圖 7-2

圖片來源：BEAMA

設計分析（Design Analysis）

在設計過程的早期階段分析能效、峰值負荷減少和脫碳措施，並使用以下方法之一將結果納入設計決策：（Analyze efficiency, peak load reduction, and decarbonization measures during the early stages of the design process and account for the results in design decision-making using at least one of the following methodologies:）

- 簡化的能源建模（Simplified energy modeling）
- 類似專案的分析（Analysis from similar projects）
- 發表數據的分析（Analysis from published data）
- 現場能源預測（Site Energy Prediction）

預測專案每年將使用的各類能源數量，並將數據提交給 USGBC。（Predict the amount of each type of energy the project will use annually in terms of site energy and submit the data to USGBC.）

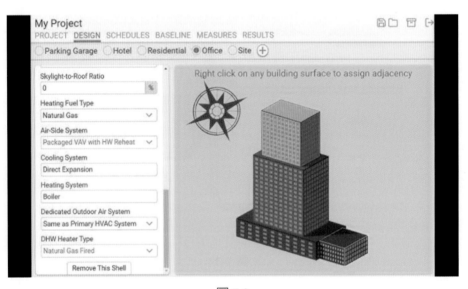

圖 7-3

圖片來源：CreateEnergy.org

碳排放預測審查（Review Carbon Projection）

根據提交的年度能源使用數據、當前電網數據和專案所在地點，USGBC 將生成專案從現在到 2050 年的碳排放「商業如常」（BAU）預

測。（Using the annual energy use data submitted, the project's current grid data, and location, USGBC will generate a "business as usual" (BAU) projection of the project's carbon emissions from energy use from the present through 2050.）

如果專案受到基於碳的建築性能標準（BPS）限制，則必須創建基於條例的 BAU，並根據該條例中定義的電氣係數和專案適用的上限進行碳排放預測。（Projects subject to a carbon-based building performance standard (BPS) must create an ordinance-specific BAU with a carbon projection based on the electrical coefficients as defined in the ordinance and an overlay showing the caps applicable to the project.）

如果超過上限，計算適用的年度罰款或費用，以及 25 年期間的累計罰款或費用。（If applicable, calculate the assessed annual fines or fees that will apply for exceeding the caps, and the cumulative fines or fees over a 25-year period.）

建築業主或其代表必須證明他們已審查 BAU 碳排放預測和費用預測。（The building owner or owner's representative shall attest that they have reviewed the BAU carbon projections and fee projection.）

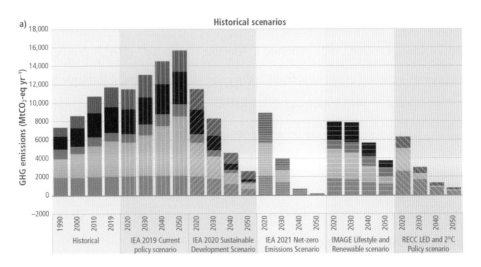

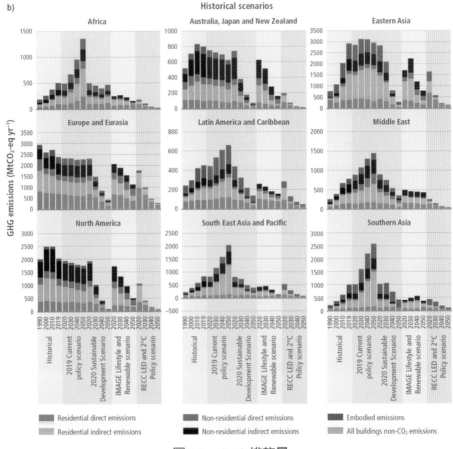

圖 7-4　GHG 排放量

圖片來源：IPCC

脫碳計畫（Decarbonization Plan）

- 制定詳細計畫，說明如何在 2050 年之前實現脫碳。（Create a plan detailing how decarbonization could be achieved by 2050.）

依國際能源總署「2050 淨零：全球能源部門路徑圖」，8% 減碳量是源於行為改變與材料效率提高，從而減少能源需求，例如減少商業目的的飛行。

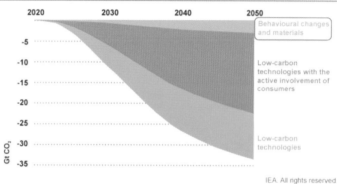

圖 7-5　國際能源總署 IEA 2050 淨零排放策略

來源：國際能源總署「2050 淨零：全球能源部門路徑圖」（Net Zero by 2050: A Roadmap for the Global Energy Sector, IEA 2021/05）https://www.iea.org/reports/net-zero-by-2050。

- 建築業主或其代表需證明已審查該脫碳計畫。（The building owner or owner's representative shall attest that they have reviewed the Decarbonization Plan.）
- 計畫應為不超過兩頁的敘述，並描述要進行的改造措施、時間表和每項措施的大致成本。（The plan shall be a narrative no more than two (2) pages in length and describe the retrofits to be made, with the approximate timeline and cost of each of the retrofit measures.）
- 描述因改造而將丟棄的設備和／或建築材料，以及將購買的新設

備。（Equipment and/or building materials that will be discarded due to the required retrofits should be described along with new equipment to be purchased.）

- 描述初始設計中納入的電氣化「準備」策略，並大致估算每項策略所避免的成本、干擾和材料浪費。（Electrification "readiness" strategies incorporated into the initial design should be described along with a rough estimate of the avoided cost, avoided disruption, and avoided materials waste afforded by each readiness measure.）

- 一些常見的準備策略包括增大電氣面板和／或服務容量、為未來負荷安裝導管、增強建築外殼或設計供暖分配系統以適應未來熱泵的較低溫度。（Some common readiness strategies include oversizing electrical panels and/or service or installing conduit for future loads, enhanced envelope, or heating distribution systems that can accommodate the lower temperatures of future heat pumps.）

圖 7-6　推動淨零綠生活的策略

來源：國家發展委員會（台灣經濟論衡）

7.2 基礎／進階能源效率（Minimum / Enhanced Energy Efficiency）

建築設計（Building Design）

策略（Strategies）

1. 能源效率（Energy Efficiency）

* 採用高效率的基礎設施（Use high-efficiency infrastructure）

基礎設施，如路燈和交通控制設備，由於其長時間使用創造顯著節能機會。（Infrastructure, such as street lighting and traffic control device, create opportunities for significant savings due to its long hours of use.）

* 進行建築物能源模擬（Use energy simulation models）

能源模擬或能源建模允許團隊看看不同的能源效率措施，看看它們是如何運作以及何種方式能提供最大的效益。（Energy simulation or energy modeling allows the team to look at the different energy efficiency measures to see how they work and which offer the greatest benefit.）

* 監測能源消耗（Monitor consumption）

如果你沒有注意能源消耗，甚至連建築物中哪裡是最消耗能源的地方都不知道。（If you are not paying attention, you even don't know how people are using building system in the way they should.）

圖 7-7　建築物能源監控系統

- 需求回應（Demand Response）

隨著後端用戶的使用電量改變，也隨之改變其電力價格，或是利用電價費率的改變，當系統可靠性不穩定時調整成浮動價格。（Changes in electric usage by end-use customers from their normal consumption patterns in response to changes in the price of electricity over time, or to incentive payments designed to induce lower electricity use at times of high wholesale market prices or when system reliability is jeopardized.）

需求回應使建築物減少他們的尖峰時段用電量，減少對電網，以及發電廠更多的應變，從而有可能避免製造新的設施成本。（Demand response allows utilities to call on buildings to decrease their electricity use during peak times, reducing the strain on the grid and the need to operate more power plant, thus potentially avoiding the costs of constructing new plants.）

- 採用高效能電器設備（Specify high-efficiency appliances）

能源之星標籤是易於識別的標誌，給予效率符合條件的設備類型，如電腦、顯示器、冰箱等產品，類似台灣的省電標章。（Energy Star label is an easily recognizable indicator of efficiency for eligible equipment types, such as computers, monitors, and refrigerators.）

圖 7-8　高效節能設備家電

2. 建築方位（Building Orientation）

• 設計適當大小的建築（Size the building appropriately）

在專案初期，專案團隊應該考量未來居住者的需求而設計建築物，但不需要超過這些需求。（Early on in the project, the project team should critically examine the needs of the future occupants and design the building to meet, but not exceed, those needs.）

找出被動式設計方式。（Identify passive design opportunities.）建築提供自然採光，並支持主動控制，可以達到顯著照明節能。（Buildings that provide access to natural lighting, and are support active controls, can achieve significant lighting energy saving.）

• 建築外殼（Building Envelope）

強調建築物外殼節能設計。（Address the building envelop.）專案在極端氣候（極熱或極冷的），隔絕效果能更顯著。（Project in extreme climate（extremely hot or cold）benefit more from higher levels of insulation.）

建築外殼節能設計問題可能導致：（Building envelope problems can lead to:）

- 很貴的冷熱空調電費帳單（high heating and cooling bills）
- 冷空氣或溫度不均（drafts or uneven temperatures）
- 空氣品質差（poor indoor air quality）
- 建築物老化（building deterioration）
- 昆蟲、噪音、氣味（insects, noise, odors）

圖 7-9　在不同的氣候區域需要有不同的外殼設計

照明設計（Lighting Design）

策略（Strategies）

1. 燈泡選項（Light Bulb Choices）：

　　選擇節能燈具（Choosing energy-efficient lighting）。

2. 燈光操控（Lighting Controls）

- 光電感知器（Photo sensors）
- 定時器（Timers）
- 人員感知器（Occupancy Sensors）
- 進階控制（Advanced Controls）

圖 7-10　各種感知器

建築用水（Water Usage）

策略（Strategies）

　　減少用水看起來可能不是一個直接節省電力的方式，但它可以減少加熱水、抽水和處理水的熱量。（Water reduction might not seem like a way to save power, but it takes energy to heat water, pump water, and treat water.）

Tankless Water Heaters（即熱式熱水器）

　　即熱式熱水器不像傳統式熱水器，會將很多熱水存放於儲桶，而是打開水龍頭才加熱產生熱水。（On-demand or tankless water heaters are products that do not constantly heat and store the hot water like conventional water heaters. Only when a hot water fixture is activated does the on-demand system heat the water and deliver it.）

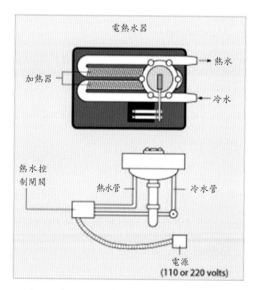

Image Source: Tankless Water Heater Guide

圖 7-11　即熱式熱水器

空調節能（Heating & Cooling）

策略（Strategies）

1. 空調設計（Design）

在高效的空調設備生命週期內，可以節省當初裝置增加之成本的許多倍。（Install high-performance mechanical often pay back many times over during equipment lifetime.）

2. 空調濾網（Air Filters）

如果過濾器具有較好的空氣穿透性，該系統將使用更少的能量來移動空氣，而在同一時間也為建築物的居住者提供更好的空氣品質。（If filters have lower resistance to air passing through them, the system will use less energy to move the air while at the same time also provide better air quality for the building occupants.）

3. 自然通風（Natural Ventilation）

全年長時間都在舒適的範圍內，就可以使用自然通風讓地區室外溫度和濕度引進室內。（Uses in areas where outdoor temperatures and humidity are within comfortable ranges for longer periods throughout the year.）

4. 地下出風系統（Underfloor Air Distribution）

放置通風管道在下方地板上，而不是天花板區域。地板系統也會改善室內空氣品質的清潔，冷氣直接傳遞到居住者的呼吸帶。（Places the ventilation ducts underneath the floor rather than in the ceiling area. Underfloor systems also improve indoor air quality as clean, conditioned air is delivered directly into occupants' breathing zones.）

5. 空調相關電器（Appliances）

能源之星電器。（ENERGY STAR appliances.）

6. 建築自動化系統（Building Automation System）

　　建築中有許多監測和調節系統，有助於減少能源使用量。所有系統皆在一個監控系統掌控之下，可以監測建築物的即時情況。（Helps reduce energy use by monitoring and regulating the many systems in a building. All systems tie into one monitoring program – more information gives those making the decisions a real time view of what is happening in a building or series of buildings.）

7.3 基本／進階建築能源系統功能驗證（Fundamental/Enhanced Commissioning of Building Energy System）

調試範圍（Commissioning Scope）

- 空調與其相關控制系統（HVAC&R and its associated control）
- 熱水系統（Domestic Hot Water supply）
- 照明（Lighting）
- 再生能源系統（Renewable Energy System）
- 綠色能源（Green Power）
- 建築物外殼（Envelop）

依循標準：ASHRAE Guideline 0 and 1.1

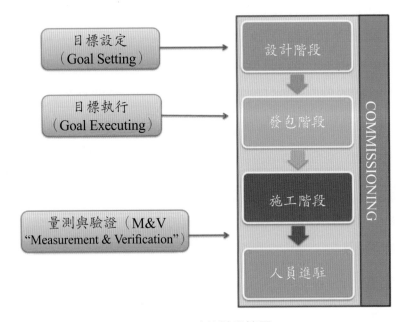

圖 7-12　功能驗證範圍

ASHRAE 90.1 規定之功能驗證範圍

1. 建築設計（Building Design）
   - 能源效率（Energy Efficiency）
   - 建築方向性（Building Orientation）
   - 建築外殼結構（Building Envelope）
2. 照明系統（Lighting Design）
   - 燈泡選擇（Light Bulb Choice）
   - 照明控制（Lighting Controls）
3. 熱水系統（Water Use）
   - 熱泵（Heat Pump）
   - 水泵（Water Pumps）

- 即熱式熱水器（Tankless Water Hesters）

4. 冷凍空調系統（Heating & Cooling）

- 冷氣過濾器（Air Filters）
- 自然通風（Natural Ventilation）
- 地板送風（Underfloor Air Distribution）
- 電器（Appliances）
- 建築設備自動化系統（Building Automation System）

5. 電力系統基本要求

6. 能源成本法（Energy Cost Budget Method）能源效率基本要求

7.4 能源量測與報告（Energy Metering and Reporting）

策略（Strategies）

　　支持能源管理實踐，並透過追蹤和報告建築的能源使用與需求，促進持續發現節能和減少溫室氣體排放的機會。（To support energy management practices and facilitate identification of ongoing opportunities for energy and greenhouse gas emissions savings by tracking and reporting building energy use and demand.）

能源監測與記錄（Energy Monitoring and Recording）

- 能源數據報告（Report Energy Data）
- 能源監測與記錄（Energy Monitoring and Recording）

　　安裝（或使用現有）設備來根據 ANSI/ASHRAE/IES 標準 90.1 監測和記錄能源使用情況。（Install (or utilize existing) devices to monitor and record energy use per ANSI/ASHRAE/IES Standard 90.1）

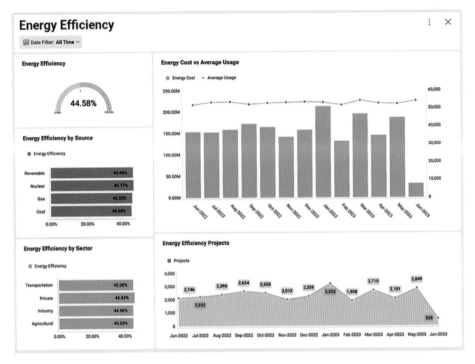

圖 7-13　能源效率示意看板（Energy efficiency dashboard）

　　安裝（或使用現有）設備來監測和記錄以下能源使用，並符合 ASHRAE 要求的電力末端使用監測和報告標準。（Install (or utilize existing) devices to monitor and record energy use for the following, meeting the same monitoring and reporting requirements as required in ASHRAE for electrical end-uses.）

現場可再生電力發電（On-site renewable electricity generation）

　　大規模翻新和符合 ASHRAE 90.1-2019 第 10.4.6 節或 90.1-2022 第 10.4.7 節例外條款的建築，必須安裝能夠至少每月監測整棟建築能源使用和建築峰值電力需求的測量設備。（Major renovations and buildings eligible for exceptions to ASHRAE 90.1-2019 Section 10.4.6 or 90.1-2022 Section

10.4.7 must install measurement devices capable of monitoring whole-building energy use for each building energy source and building peak electricity demand at least monthly.）

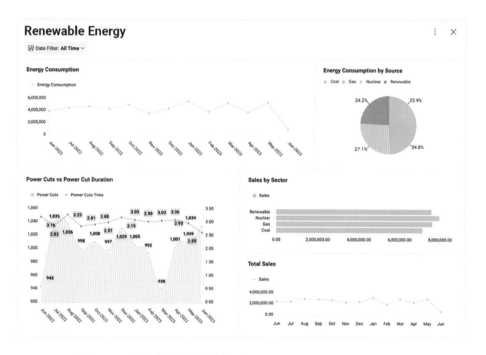

圖 7-14 再生能源示意看板（Renewable energy dashboard）

能源數據報告（Report Energy Data）

承諾至少每年向 USGBC 報告以下數據：（Commit to reporting the following data to USGBC at least annually:）

- 連續 12 個月的每個能源來源的每月總能耗數據（monthly energy data for 12 consecutive months of total energy consumption for each energy source）
- 現場可再生能源發電數據（on-site renewable energy generation）

- 峰值電力需求數據（peak electrical demand）
- 此承諾必須持續五年，或直到建築物更換業主或租戶爲止（This commitment must carry forward for five years or until the building changes ownership or lessee）

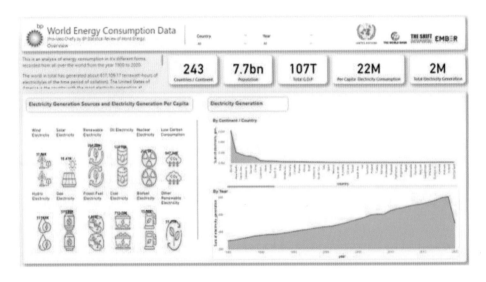

圖 7-15　能源數據報告（Report Energy Data）
圖片來源：WORLD ENERGY CONSUMPTION REPORT

7.5 基礎／進階冷媒管理（Fundamental／Enhanced Refrigerant Management）

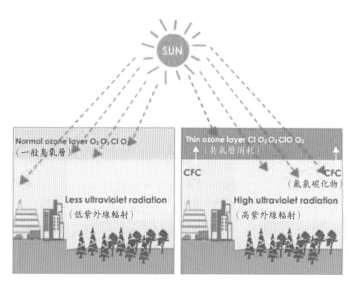

圖 7-16 臭氧層破壞示意圖

臭氧層破壞潛力（Ozone Depletion Potential (ODP)）

所述物質引起的臭氧消耗的量。（The amount of ozone depletion caused by a substance.）

全球溫暖化潛力（Global Warming Potential (GWP)）

所述物質引起的全球變暖的量。（The amount of global warming caused by a substance.）

冷媒管理（Refrigerant Management）

1. 氟氯碳化物（CFC: Chloro-Fluoro-Carbon）

　　不可使用，既有建築內的空調設備需提出汰換計畫。CFC 產品到 2010 年完全禁止生產。

2. 氫氟氯碳（HCFC: Hydro-Chloro-Fluoro-Carbon）

　　將來需逐步汰換，使用 HCFC 的設備只能生產至 2020 年，2030 年以後完全禁止生產。

3. 氫氟碳（HFC: Hydro-Fluoro-Carbon）

　　可繼續使用。

4. 天然冷煤（Natural refrigerant）

　　可繼續使用，包括：丙烷（Propane）、二氧化碳（CO_2）、氨（NH_3）、水（Water）、空氣（Air）。

	ODP	GWP	
氟氯碳化物（Chlorofluorocarbons）			
CFC-12	1.0	10,720	不好
CFC-11	1.0	4,680	（Worst）
氫氟氯碳化物（Hydrochlorofluorocarbons）			
HCFC-22	0.04	1,780	
HCFC-123	0.02	76	
氫氟碳化物（Hydrofluorocabons）	-0	12,240	
HFC-23	-0	1,320	
HFC-134a			
自然冷媒──碳氫化合物（Natural Refrigerants-Hydrocarbons (HC)）	0	3.0	
乙烷（Ethane）			
丙烷（Propane）	0	3.0	
丁烷（Butane）	0	3.0	

自然冷媒——其他（Natural Refrigerants-other）			
Carbon Dioxide (CO_2)	0	1.0	
Ammonia (NH_3)	0	0	
Water (H_2O)	0	0	
Air	0	0	好（Best）

- 氟氯碳化物（CFC (Chloro-Fluoro-Carbon)）
  - 新建案完全禁止使用。（Zero use of chlorofluorocarbon (CFC)-based refrigerants in new base building heating, ventilating, air conditioning and refrigeration (HVAC&R) systems.）
  - 既有建築：空調設備需進行效益評估與汰換計畫。（Existing base building HVAC equipment, complete a comprehensive CFC phase-out conversion prior to project completion.）
- 不得安裝含有 CFC、HCFC 或 Halons（海龍）等會破壞臭氧層的消防系統。（Do not operate or install fire suppression systems that contain ozone-depleting substances such as CFCs, hydrochlorofluorocarbons (HCFCs) or halons.）

適用標準與規範（Standards that Support LEED Credit）

ASHRAE 90.1（美國冷凍空調協會規範 90.1）	California T-24	Emergy Star Program（能源之星計畫）	IPMVP（國際性能量測與確認協定）	Green-e Certification
Energy Standards（能源規範）	Parallel to ASHRAE 90.1（與 ASHRAE 平行的加州能源規範）	Online Energy Performance tool（線上能源計畫工具）	Verify the energy performance（確主耗能量化數據）	Renewable energy（可再生能源）

圖 7-17　一張圖了解能源與大氣

圖片來源：USGBC

7.6 電氣化（Electrification）

策略（Strategies）

鼓勵建築物的設計不依賴現場燃料燃燒，從而改善室內外空氣品質，並隨著電網脫碳，實現低碳運營。（To encourage buildings to be designed so they do not depend on burning fuel on-site, leading to better indoor and outdoor air quality and to low carbon operations as the grid decarbonizes.）

無現場燃燒（No On-Site Combustion）

設計並運營專案，從進駐開始不進行現場燃燒，緊急支援系統除外。（Design and operate the project from start-up with no on-site combustion except for emergency support systems.）

　　空間暖氣和熱水供暖的設備加權平均效率必須至少為 1.8 COP。
（Combined weighted average equipment efficiency for space heating and service water heating must be at least 1.8 COP.）

　　以下設備可排除在 COP 計算之外：（The following equipment may be
excluded from the COP determination:）

- 氣候區 0 至 2 的空間加熱設備。（Space heating equipment in climate
  zones 0 through 2.）

- 專為在室外空氣乾球溫度（DBT）低於 20°F（–6.5℃）時運行的輔
  助加熱設備。（Supplemental or auxiliary heating equipment designed
  only for operation at outside air dry-bulb temperatures (DBT) below
  20°F (–6.5℃).）

- 非住宅空間中的服務熱水設備，符合 ASHRAE 90.1-2022 第
  11.5.2.3.3 節的即時熱水器標準，且無例外。（SWH equipment in
  non-residential spaces complying with the point-of-use water heater
  criteria in ASHRAE 90.1-2022 Section 11.5.2.3.3, W05, without exceptions.）

除低溫運行外，無現場燃燒（No On-Site Combustion Except at Low Temperatures）

路徑 1：空間加熱（Path 1. Space Heating）

　　設計空間供暖系統，在室外空氣（OSA）溫度高於 20°F（–6.5℃）時，
能夠在無現場燃燒的情況下運行。氣候區 3 及以上的專案，其加權平均空
間加熱設備效率必須至少為 1.8 COP。（Design space heating to be capable
of operating without on-site combustion at outside air (OSA) temperatures above

20°F (–6.5℃). Projects in climate zones 3 and above must have a weighted average space heating equipment efficiency of at least 1.8 COP.）

路徑 2：熱水供暖加熱（Path 2. Service Water Heating）

設計服務熱水供暖系統，能夠在室外空氣溫度高於 20°F（–6.5℃）時無現場燃燒運行。服務熱水加熱能力超過 34,000 Btu/h（10 kW）的專案，其加權平均設備效率必須至少為 1.8 COP 或太陽能熱水比例至少達到 0.4。（Design service water heating systems to be capable of operating without on-site combustion at outside air temperatures above 20°F (–6.5℃). Projects with total service water heating capacity exceeding 34,000 Btu/h (10 kW) must have a weighted average service hot water equipment efficiency of at least 1.8 COP OR domestic hot water solar fraction of at least 0.4.）

路徑 3：烹飪及其他作業負荷（Path 3. Cooking and Other Process Loads）

設計烹飪、洗衣、作業設備及現場發電系統（不包括緊急支援系統）能夠在無現場燃燒的情況下運行。（Design cooking, laundry, process equipment, and on-site power generation except emergency support systems to be capable of operating without on-site combustion.）

設備效率（Equipment efficiency for Options 1 and 2）

- 按額定條件計算設備效率。對於具有多個額定條件的設備，使用最接近 17°F（–9℃）外界乾球溫度、32°F（0℃）進水溫度或 44°F（6℃）加熱源出水溫度的額定值。（Equipment efficiencies at rated conditions. For equipment with multiple rated conditions, use the rating closest to 17°F (–9℃) OA db, 32°F (0℃) entering liquid temperature, or 44°F (6℃) heating source leaving liquid temperature.）

- 使用能源模擬計算年度平均 COP。（Annual average COP calculated with an energy simulation.）

燃料電池（Fuel Cells）

使用化石燃料的燃料電池不符合得分條件。（Fuel cells using fossil fuel are ineligible for credit.）

名詞小教室

燃料電池

燃料電池，是一種將化學能轉化為電能的電化學裝置。通過電解液，電池內部的燃料與氧化劑（例如氧氣）發生化學反應，從而產生電力。這個過程基本上是水電解的逆反應（$H_2 + 1/2O_2 \rightarrow H_2O + $ 電力）。

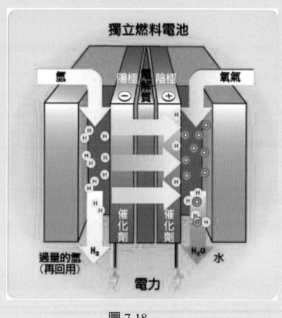

圖 7-18

7.7 降低尖峰熱負荷（Reduce Peak Thermal Loads）

目標（Intent）

最小化電網資源的需求並提高建築的韌性。（To minimize demand on grid resources and improve the resiliency of buildings.）

氣流滲透和平衡通風（Infiltration and Balanced Ventilation）

符合以下兩個要求：（Comply with both of the following:）

- 平衡通風（**Balanced Ventilation**）：設計供氣和排氣流量在 10% 內，並在調試範圍內包括測試、調整和平衡（TAB）報告以證明平衡通風。（Design the supply and exhaust airflows within 10% of each other and include a Test, Adjusting, and Balance (TAB) report demonstrating balanced ventilation in the commissioning scope.）

- 氣流滲透（**Infiltration**）：使用氣密性測試，證明建築外殼的氣密性等於或低於要求。（Use an air leakage test to demonstrate a measured air leakage of the building envelope less than the caps.）

通風能量回收（Ventilation Energy Recovery）

每個供應室外空氣的風機系統必須配備一個能量或熱回收系統，最小 70% 的焓回收率或最小 75% 的顯熱回收率。（Each fan system supplying outdoor air must have an energy or heat recovery system with a minimum 70% enthalpy recovery ratio or a minimum of 75% sensible heat recovery ratio.）

HOME VENTILATION SYSTEM

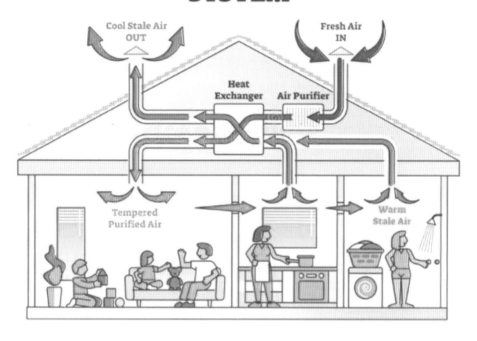

HEAT RECOVERY VENTILATOR

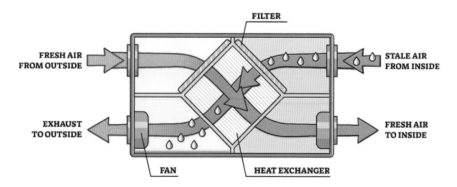

圖 7-19 全熱交換機

熱橋效應（Thermal Bridging）

遵循 ASHRAE 90.1-2022 第 5.5.5(a) 節的規定，無例外地滿足所有適用的要求。（Comply with the prescriptive thermal bridging requirements of ASHRAE 90.1-2022, Section 5.5.5(a), without applying exceptions.）

峰值熱負荷減少（Peak Thermal Load Reductions）

包含通風負荷以計算峰值同時負荷。（Ventilation loads must be included in the determination of peak coincident loads.）

使用氣密性測試結果來計算峰值負荷，並展示符合的平衡通風要求。（Measure building envelope air leakage and use it to calculate peak loads for Path 1, Path 2 (envelope), and Path 3.）

路徑 1：峰值負荷強度（Path 1. Peak Load Intensity）

限制每平方英尺處理樓板面積的峰值加熱和冷卻負荷之和低於或等於表 2 中的閾值。（Limit the sum of peak heating load and peak cooling load per unit of treated floor area to be less than or equal to the thresholds specified in Table 2.）

路徑 2：ASHRAE 90.1 折衷方法（Path 2. ASHRAE 90.1 Trade-Off Methods）

根據 ASHRAE 90.1-2022 的建築外殼折衷選項，展示比基準外殼性能係數更好的百分比改善。（Demonstrate a percent improvement in the sum of system peak heating loads and system peak cooling loads associated with the proposed envelope performance factor compared to the base envelope performance factor determined in accordance with ASHRAE 90.1-2022.）

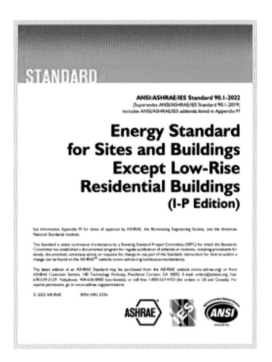

圖 7-20　Ashrae 90.1 建築能源標準用戶手冊（ASHRAE Standard 90.1）

名詞小教室

　　1975 年，ASHRAE 90-75：新建築設計中的節能（ASHRAE 90-75: Energy Conservation in New Building Design）被發布，成為美國第一部全國性能源法規。隨後，在 1989 年版中，ASHRAE 90 分為 ASHRAE 標準 90.1：非低層住宅建築的能源標準（ASHRAE Standard 90.1: Energy Standard for Buildings Except Low-Rise Residential Buildings）和 ASHRAE 90.2：低層住宅建築的節能設計（ASHRAE 90.2: Energy-Efficient Design of Low-Rise Residential Buildings）。這一分離確立了標準 90.1 作為 ASHRAE 對商業建築能源效率如何進行監管和評估的指導答案。（In 1975, ASHRAE 90-75: Energy Conservation in New Building Design was published as the first national energy code. Later, as part of the 1989 edition, ASHRAE 90 split into ASHRAE Standard 90.1: Energy Standard for Buildings Except Low-Rise Residential Buildings and ASHRAE 90.2: Energy-Efficient Design of Low-Rise Residential Buildings. This split solidified Standard 90.1 as ASHRAE's answer to how to regulate and evaluate energy efficiency in commercial buildings.）

根據太平洋西北國家實驗室的研究，標準 90.1 的後續版本，直至 ASHRAE 90.1-2016，已將依據該標準設計的典型建築的能源消耗減少了 50% 以上。這一進展顯示於圖 1 中。ASHRAE 90.1-2019 是最新發布的版本，但尚未納入上述研究或被廣泛採用。（Successive versions of Standard 90.1, up to ASHRAE 90.1-2016, have reduced the energy consumed by a typical building designed to the standard by over 50% according to research done by Pacific Northwest National Laboratory. This progress is shown in Figure 1. ASHRAE 90.1-2019 is published as the most recent version, but has not been analyzed as part of the research mentioned or extensively adopted.）

標準 90.1 的逐步進展，使建築設計脫離了對能源使用沒有任何法規要求的時代，當時如牆體隔熱等重要的建築特徵完全取決於設計和施工團隊的個人動機。（The cumulative and methodical progress of Standard 90.1 has taken building design out of the era when no code requirements were placed on energy usage, and important building characteristics such as the inclusion of wall insulation were left up to the individual motivations of the design and construction teams.）

路徑 3：能源模擬（Path 3. Energy Simulation）

　　根據 ASHRAE 90.1-2019 或後續版本的附錄 G，展示性能指數（PI）計算，並將所有「成本」參考替換爲峰值加熱和冷卻負荷之和。（Demonstrate a Performance Index (PI) calculated per ASHRAE 90.1-2019 or later Normative Appendix G, replacing all references to "cost" with "the sum of building peak coincident heating loads and building peak coincident cooling loads."）

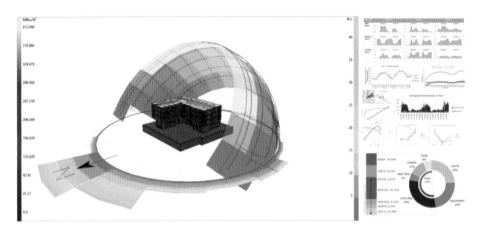

圖 7-21　建築能源模擬（Energy Simulation）

來源：Egreenideas.com

　　建築在其生命週期中所消耗的能源，對環境的影響超過設計、施工或運營的任何其他方面。它也是建築物業主的最大成本之一。因此在設計過程的早期對新建築進行能源性能建模至關重要：這就是設計高性能建築的方式。能源模型幫助專案團隊了解建築的建築設計、機械設計和電氣設計決策對預測能源性能的生命週期影響。（The energy consumed by a building throughout its lifetime has the biggest impact on the environment of any other aspect of design, construction or operations. It is also one of the biggest costs to a building owner.That it is critically important to model the energy performance of a new building early in the design process: this is how high-performance buildings are born. Energy Modeling helps a project team understand the life-cycle impacts of architectural, mechanical and electrical design decisions on the projected energy performance of the building.）

能源模型實質上讓團隊成員可以比較不同的設計選項，考慮初始成本、運營成本和更換成本。最終，能源模型賦予業主建造出性能優於同類建築的能力，這可能是市場上的競爭優勢。（Modeling essentially lets team members compare different design options by considering first cost, operating costs, and replacement costs. Ultimately, Energy Modeling gives the owner the ability to create a building that performs better than similar buildings, which can be a competitive advantage in the market.）

7.8 可再生能源（Renewable Energy）

策略（Strategies）

1. 基地內產生的再生能源（Generate on-site renewable energy）使用直射太陽光線使太陽能（PV）面板利用太陽能系統發電，或太陽能熱水系統產生熱水。（Project with access direct sunlight can generate electricity using photovoltaic（PV）panels and hot water using solar hot-water systems.）

2. 採購基地外的再生能源（Purchase off-site renewable energy）工程專案可以從公用事業或私人電廠收購綠色電力。（Project can purchase green power from their utility.）

3. 使用免費的能源（Use free energy）許多建築基地都有潛力透過被動系統去滿足大多數人的需求，如採光、自然通風，以及使用該建築物的熱質量吸收。（Many building sites have the potential to satisfy the majority of human needs through passive systems such as daylighting, natural ventilation,

and using the building mass as a thermal system.）

再生能源（Renewable Energy）

1. 現場可再生能源（On-Site Renewable Energy）

　　　(1) 太陽熱能（Solar Thermal Systems）

　　　(2) 太陽能光電系統（Photovoltaic Systems）

　　　(3) 風能（Wind Energy）

　　　(4) 生物質能（Biomass）

圖 7-22　太陽能發電板

2. 再生能源（Renewable Energy）

　(1) 再生能源（Renewable Energy）

　　• 太陽能光電系統（Photovoltaic system）

圖 7-23　太陽能光電板

- 風能（Wind energy）

圖 7-24　風力發電

- 太陽熱能（Solar thermal energy）

圖 7-25　太陽能熱水器

- 生質燃料發電（Bio-fuel-based electrical system）
- 地熱能（Geothermal heating）
- 地熱電能（Geothermal electrical）
- 低環境影響之水力發電（Low-impact hydro-electric power system）

圖 7-26　水力發電

- 潮汐發電（Wave and tidal power）

(2) 生質燃料包含（Bio-fuel）

- 未處理過的廢木，包含鋸木廠的木屑（Un-treated wood waste, include mill reside）
- 農作物或農作廢料（Agricultural crops or waste）
- 動物排泄物或其他有機廢料（Animal waste and other organic waste）
- 掩埋場的沼氣（Landfill gas）

(3) 不符合 LEED 再生能源資格的包含：

- 建築物外殼元素設計（Architectural features）
- 被動太陽能措施（Passive solar strategies）
- 晝光控制（Day lighting strategies）
- 地熱交換系統（Geo-exchanger system）

3. Free Energy

地源熱泵（Geothermal Heat Pumps）

注意：LEED 認定 Geothermal 不算 Renewable Energy。

透過管線使空氣與地下土地進行熱交換，以減輕空調負荷。

圖 7-27　地源熱泵

綠色能源（Green Power）

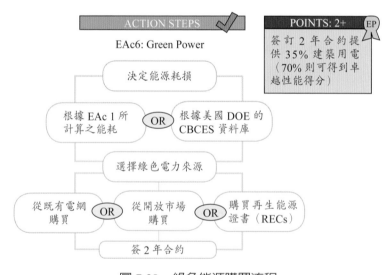

圖 7-28　綠色能源購買流程

策略（Strategies）

圖 7-29　再生能源

1. 非現場可再生能源（Off-Site Renewable Energy）

- 風能（Wind Energy）
- 太陽能（Solar Energy）
- 水能（Hydro Energy）
- 生物燃料（Biofuel）
- 潮汐發電系統（Wave and Tidal Power Systems）

可再生能源證書（REC）（Renewable Energy Certificate(REC)）

　　REC 證書對於環境、社會有著正面影響，REC 是透過可再生資源產生的電力所頒發的購買證明。（A REC represents the environmental, social, and other positive attributes of power generated by renewable resources.）

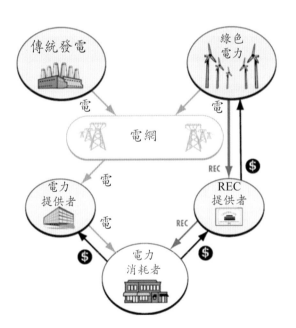

圖 7-30　REC 運作模式

為了獲得 LEED 信用，區域經濟共同體必須是 Green-c 認證

（To caen LEED credit, EECs must be Green-e Certified）

在美國還有一種專門的可再生能源證書 REC，即 Renewable Energy Certificates。由於可再生能源證書 REC 被定義為一種可交易的、無形的能源商品，因此電力企業可以在可再生能源證書交易市場，從可再生能源發電企業手中購買其多出來的可再生能源證書。與碳交易的全球性不同的是，RECs 只能在本國（地區）內交易，不能跨國交易。（In the United States there is a special renewable energy certificates REC, namely Renewable Energy Certificates. Since renewable energy certificates REC is defined as a tradable commodity invisible energy, so power companies can renewable energy certificate trading market to buy out its rich renewable energy certificates from renewable energy enterprises in the hands. Global carbon trading and the difference is, RECs can only be traded in within the country (region), not cross-border transactions.）

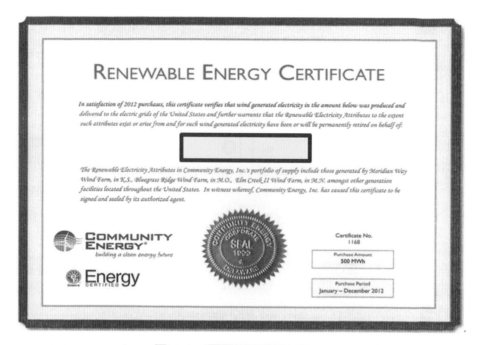

圖 7-31　可再生能源證書樣本

影響（Impacts）

1. 問題（Issue）

除了優良的設計以外，建築物節能眞正的挑戰在人員進駐之後。要達到最佳的能源效率，只有靠使用者的用心維護與操作，才能達到當初設計的要求。（To reduce the environmental and economic harms of excessive energy use by achieving a minimum level of energy efficiency for the building and its systems.）

2. 經濟（Economy）

建築物眞正的節能成效，更要建築物進駐後的正確操作與維護才能顯現。要降低建築物操作成本，也要靠正確的操作與維護

3. 進行預防性的保養維護（Conduct preventive maintenance）

定期維護有助於確保系統停留在巔峰狀態。（Periodic maintenance helps ensure that system stay in peak condition.）

4. 獎勵住戶和承租戶（Create incentives for occupants and tenants）

該獎勵可以幫助鼓勵居民，當他們在不使用時關掉電燈和設備。（The incentive can help encourage occupants to turn off lights and equipment when they are not in use.）

5. 符合業主需求（Adhere to owner's project requirements）

當評估設計方案和功能測試系統一旦建立時，業主工程專案要求（OPR）應被用來作爲專案的基準。（The owner's project requirement (OPR) should be used as a benchmark for achievement when evaluating design options and functionally testing the systems once constructed.）

6. 進行員工訓練（Provide staff training）

人員培訓以確保建築管理者了解如何維護和操作複雜的系統。（Staff training resources ensure that building operator understand how to maintain and operate complex systems.）

7.9 互動式電網（Grid-Interactive）

策略（Strategies）

提高電力韌性，並使建築成為應對電網變化、支持電網脫碳、穩定性和電力可負擔性的積極合作夥伴。（To enhance power resiliency and position buildings as active partners contributing to grid decarbonization, reliability, and power affordability through integrated management of building loads in response to variable grid conditions.）

能源儲存（Energy Storage）

提供現場電力儲存和／或熱能儲存。包括能夠在非尖峰期或電網碳排放強度低時儲存電力或熱能的自動負載管理控制，並在尖峰期或碳排放強度高時使用儲存的能量。（Provide on-site electric storage and/or thermal storage. Include automatic load management controls capable of storing the electric or thermal energy during off-peak periods or periods with low grid carbon intensity and using stored energy during on-peak periods or periods of high grid carbon intensity.）

需求回應計畫（Demand Response Program）

與合格的需求回應（DR）計畫提供商簽訂至少一年的合約，並有多年的續約意向。現場電力發電和燃料燃燒不能用於需求側管理標準。（Enroll in a minimum one-year demand response (DR) contract with a qualified DR program provider, with the intention of multiyear renewal. On-site electricity generation and fuel combustion cannot be used to meet the demand-side management criteria.）

自動化需求側管理（Automated Demand Side Management）

現場電力發電和燃料燃燒不能用於需求側管理標準。（On-site electricity generation and fuel combustion cannot be used to meet the demand-side management criteria.）

路徑 1：系統級控制（Path 1. System-Level Controls）

提供自動需求回應（ADR）控制系統，至少控制以下兩個系統：（Provide automated demand response (ADR) controls for at least two of the following systems installed within the project scope of work:）

- HVAC 系統（50% 的額定容量）（HVAC systems (50% of rated capacity)）
- 照明系統（50% 的電力）（Lighting systems (50% of power)）
- 自動插座控制（Automatic receptacle controls）
- 服務熱水加熱（90% 的容量）（Service water heating (90% of capacity)）
- 電動車供電設備（EVSE）（Electric vehicle supply equipment (EVSE)）

路徑 2：建築自動化系統（Path 2. Building Automation System）

制定一個計畫，至少減少專案尖峰電力需求的 10%，持續至少 1 小時，並考慮多夏季尖峰。系統需能根據電網壓力或碳排放觸發信號，自動減少電力需求。（Develop a plan for shedding at least 10% of the project's peak electricity demand for a minimum of one hour. The plan must address both winter and summer peaks considering electrified grid projections. Have in place a control system that automatically sheds electricity demand in response to triggers denoting strain on the grid or high grid emissions.）

電力韌性（Power Resiliency）

辨識需要持續運行的關鍵設備，並設計建築使其能夠與電網獨立，依靠現場的可再生能源和儲能系統，為關鍵負載項目提供至少三天的電力。（Identify critical equipment that requires continuous operation. Design the project to be able to island and operate independently from the grid to power the critical loads with the project's onsite renewable and energy storage systems for at least three days.）

名詞解釋（Glossary）

- 功能驗證（Commissioning (Cx)）
  建築和所有的系統及組件的計畫、設計、安裝、測試、操作和維護，以滿足業主的專案要求之過程的核實和記錄（The process of verifying and documenting that a building and all of its systems and assemblies are planned, designed, installed, tested, operated, and maintained to meet the owner's project requirements.）

- 能源之星標準評估系統（ENERGY STAR® Rating）
  建築物之能源性能的衡量標準與具有相似特徵的建築相比，透過 ENERGYSTAR® 管理系統的使用。50 分代表平均建築性能。（A measure of a building's energy performance compared with buildings with similar characteristics, as determined by use of the ENERGY STAR® Portfolio Manager. A score of 50 represents average building performance.）

- 地熱能源（Geothermal energy）
  透過從地球內轉換熱水或蒸汽產生電。（Electricity generated by converting hot water or steam from within Earth.）

- 海龍（Halons）

  在滅火系統和滅火器使用的物質，會消耗平流層臭氧層。（Substances used in fire-suppression systems and fire extinguishers. These substances deplete the stratospheric ozone layer.）

- 可再生能源證書（RECs）（Renewable energy certificates (RECs)）

  再生能源證書代表其電力來自再生能源系統，而發展出可交易的模式；再生能源證書的費用直接給予再生能源電廠，鼓勵了綠色能源的發展，因此 LEED 希望用戶購買綠色電力，並提出 REC 來獲得分數。

  （Tradable environmental commodities representing proof that a unit of electricity was generated from a renewable energy resource; RECs are sold separately from the electricity itself and thus allow the purchase of green power by a user of conventionally generated electricity.）

題目（Practice Questions）

1. 可再生能源證書（RECs）代表什麼？（To what do renewable energy certificates (RECs) refer?）

   a. 基地內的太陽能光電系統（On-site photovoltaic systems）

   b. 基地外的再生能源交易（Off-site renewable energy purchases）

   c. 市場上產生的風力發電（Market-generated wind power）

   d. 存在公用事業供應商（Stock in utility providers）

   RECs 是一種可交易的能源商品，證明每單位（1 兆）電力是由再生能源產生，RECs 可以與其產生的電力本身分開交易。（RECs are tradable environmental commodities representing proof that a unit of electricity was generated from a renewable energy source. RECs are sold separately from the electricity itself.）

2. 其中最具成本效益，可確保最佳的持續能源性能是＿＿＿？（One of the most cost-effective ways to ensure optimal ongoing energy performance is to ＿?）

　　a. 功能驗證系統（Commission building systems）

　　b. 升級機械系統（Upgrade mechanical systems）

　　c. 安裝可再生能源系統（Install renewable energy systems）

　　d. 維護樹木和美化功能（Maintain trees and landscaping features）

勞倫斯伯克利國家實驗室的研究發現，調試既有建築為每平方英尺 $0.27 元，平均的回收期為 0.7 年。對於新建築研究發現，基於節能投資回收期不得不為 4.8 年，平均成本為每平方英尺 $1 元。（The Lawrence Berkeley National Laboratory study found that commissioning for existing buildings has a median cost of $0.27 per square foot and an average simple payback of 0.7 years. For new construction, the study found that the median cost was $1 per square foot and had a payback of 4.8 years based on energy savings alone.）

3. 辦公設備、廚房炊事、自動扶梯所使用的能源被稱為？（Energy use associated with office equipment, kitchen cooking, and escalators is known as?）

　　a. 日常能源（Regulated energy）

　　b. 除外能源（Exempt energy）

　　c. 作業能源（Process energy）

　　d. 二次能源（Secondary energy）

作業能源被定義為用於運行辦公設備、電腦、電梯和自動扶梯、廚房做飯和製冷設備、洗衣房洗滌和乾燥裝置的能源。（Process energy is defined as energy used to run office equipment, computers, elevators and escalators, kitchen cooking and refrigeration units, laundry washing and drying units, lighting that is exempt from the lighting power allowance, and miscellaneous items.）

4. 下列何者被認為是可再生能源（選擇兩個）？（Which are considered renewable energy sources (select two)?）

a. 原子核（Nuclear）

b. 太陽能（Solar）

c. 潮汐（Wave）

d. 天然氣（Natural gas）

可再生能源來自那些不會耗竭的能源，例如包含太陽能、風力、水力，再加上地熱及潮汐系統的能源。（Renewable energy comes from sources that are not depleted by use. Examples include energy from the sun, wind, and small (low-impact) hydropower, plus geothermal energy and wave and tidal systems.）

5. 哪個政府部門管理能源之星計畫？（選擇一項）（Which government agency manage Energy Star Program? (select one)）

a. 白宮（White House）

b. 能源部門（Energy Department）

c. 環保署（EPA）

d. 農業局（FDA）

e. 美國農業部（USDA）

ENERGY STAR 是美國環境保護署（EPA）發起的計畫，幫助企業和個人節省資金，並透過卓越的效能保護我們的環境。（ENERGY STAR is a U.S. Environmental Protection Agency (EPA) voluntary program that helps businesses and individuals save money and protect our climate through superior energy efficiency.）

6. 關於氫氟氯碳（HCFC）哪種說法是正確的？（Which statement is true regarding HydroCholoFluoroCarbons (HCFC)?）

　　a. 蒙特利爾議定書禁止氫氟氯碳冷煤（HCFCs are refrigerants banned by the Montreal Protocol）

　　b. 氫氟氯碳是一種天然冷煤（HCFCs are a type of natural refrigerant）

　　c. LEED 要求氫氟氯碳為零（LEED requires zero of HCFCs LEED）

　　d. 氫氟氯碳有比氟氯碳化物低的臭氧消耗潛能（HCFCs have a low ozone depletion potential of CFCs）

氫氟氯碳的臭氧消耗潛能比氟氯碳化物低。氫氟氯碳還沒有被蒙特利爾議定書禁止。它不是天然冷煤。LEED 要求不能有 CFC，但 HCFC 還可以存在。（HCFCs have a low ozone depletion potential of CFCs. HCFC has not been banned by Montreal Protocol. It is not a natural refrigerant. LEED require zero CFC, not HCFC.）

7. 一個冷凍庫在設計階段可以採取下列哪些行動降低其電力需求？（A refrigerated warehouse project in the design phase could reduce its electricity demand by taking which of the following action?）

　　a. 購買碳抵消應對溫室氣體排放量。（Purchasing carbon offsets to counter GHG emissions.）

　　b. 購買來自風力發電場所有的電力（Purchasing all electricity from a wind farm）

　　c. 需求回應的加入（Enrolling in a demand response program）

　　d. 採購氟氯碳化物的可再生能源證書（Purchasing Renewable Energy Credits (RECs)of CFCs）

需求回應程序可以在能源要求嚴格的專案使用：數據中心、冷藏庫或在夏天華氏 110 度以上建築地區（A Demand response program can work for demanding energy projects- data center, refrigeration, or full occupied building in areas that hit 110F degrees in the summers）

第八章　材料與資源（Materials and Resources）

學習目標（Learning Objectives）

- 減少廢棄物 - 廢棄物管理（Reducing Waste）
  - 固體廢棄物管理的階層（Solid Waste management Hierarchy）
  - 可回收物的儲存和收集（Storage & Collection of Recyclables）
  - 營建與拆除廢棄物管理計畫（Construction and Demolition Waste Mgmt Planning）
- 建築生命週期的影響降低（Building life- cycle impact reduction）
  - 環保材料（Environmentally preferable materials）
  - 環境產品聲明（Environmental Product Declarations）
  - 原材料的採購（Sourcing of Raw Materials）
  - 永續性採購計畫（Sustainable Purchasing Program）

減少廢棄物（Reducing Waste）

固體廢棄物管理的階層（Solid Waste management Hierarchy）

傳統廢物管理階層 V.S. 潛能開發。（Traditional Waste Hierarchy V.S. Potential Development.）

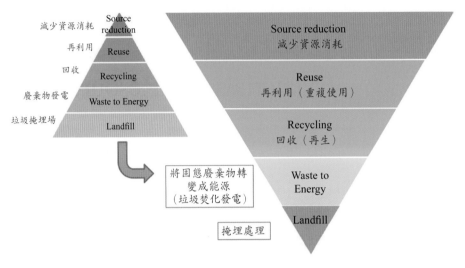

圖 8-1 廢棄物管理階層圖

8.1 零廢棄營運計畫（Planning for Zero Waste Operations）

　　從設計、製造、購買或使用的材料來達到減少產生的量或垃圾之毒性的做法。（It's the practice of designing, manufacturing, purchasing, or using materials in ways that reduce the amount or toxicity of trash created.）

　　定義：減少帶進建築之不必要的材料數量，以產生更少的廢棄物，例如：購買少量包裝的產品就是一種減少資源消耗的策略。（Definition: A decrease in the amount of unnecessary material brought into a building in order to produce less waste. For example, purchasing products with less packaging is a source reduction strategy.）

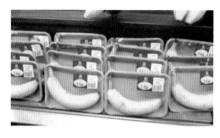

圖 8-2　多餘浪費的包裝

鼓勵採用創新建築策略，如預製組件和訂製適當尺寸的建築材料，從而最大限度地降低材料的更換週期值和低效率。（Source reduction encourages the use of innovative construction strategies, such as **prefabrication** and **designing to dimensional** construction materials, thereby minimizing material cutoffs and inefficiencies.）

模組化建築
（Modular and prefabricated buildings）

結構式絕緣墙板
（Structural Insuated Panel (SIP)）

圖 8-3　模組化的構造

將固態廢棄物轉變成能源（Waste to Energy）

將固態廢棄物轉變成能源（垃圾焚化發電）（Waste-to-Energy）

將無法回收的廢棄物轉化為有用的熱、電、燃料，或透過其他的方

式，例如：焚化、氣化、厭氧消化、垃圾掩埋沼氣回收等。（It's the conversion of non-recyclable waste materials into useable heat, electricity, or fuel through a variety of processes, including combustion, gasification, anaerobic digestion, and landfill gas (LFG) recovery.）

- 降低對石化能源的依賴（Offsetting the need for energy from fossil sources）
- 減少掩埋沼氣（甲烷）（Reduces methane generation from landfills）

在 1996 年至 2007 年，美國已經沒有再建造新的焚化爐，其主要原因有：(1) 經濟因素：隨著大型的低成本地區性垃圾堆填區的增加，現今電力的價格相對較低。(2) 稅收政策：美國在 1990 年至 2004 年廢除了由廢棄物發電的發電廠的稅收抵免。（No new incinerators were built between 1996 and 2007.The main reasons for lack of activity have been: (1)Economics. With the increase in the number of large inexpensive regional landfills and, up until recently, the relatively low price of electricity, incinerators were not able to compete for the 'fuel', i.e., waste in the U.S.(2)Tax policies. Tax credits for plants producing electricity from waste were rescinded in the U.S. between 1990 and 2004.）

近年來在美國和加拿大，對焚燒垃圾和其他垃圾轉換為能源的技術又燃起了新的興趣。2004 年，垃圾焚燒在美國獲得可再生能源生產的稅收抵免資格。增加現有工廠容量的專案正在進行中，並且市府再一次評估建設焚燒廠，而不是選擇繼續採用堆填區的方式處理城市垃圾。Source: Wikipedia（There has been renewed interest in incineration and other waste-to-energy technologies in the U.S. and Canada. In the U.S., incineration was granted qualification for renewable energy production tax credits in 2004.Projects to add capacity to existing plants are underway, and municipalities are once again evaluating the option of building incineration plants rather than continue landfilling municipal wastes.）

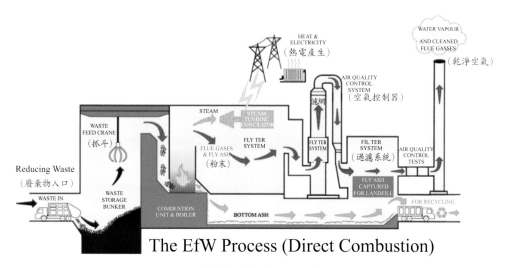

The EfW Process (Direct Combustion)

圖 8-4　垃圾焚化燃燒過程示意圖

8.2 評估並量化隱含碳（MR Prerequisite: Assess and Quantify Embodied Carbon）

策略（Strategies）

　　評估並量化專案結構、外殼及硬鋪面材料的隱含碳影響，並找出最大的減排機會。（To assess and quantify the embodied carbon impacts of the structure, enclosure, and hardscape of a project and identify the largest opportunities for reductions on the project.）

隱含碳（Embodied Carbon）

　　量化專案結構、外殼和硬鋪面材料的隱含碳影響（全球變暖潛勢

GWP）。至少包括：瀝青、混凝土、磚石、結構鋼、隔熱材料、鋁擠壓件、結構木材與複合材料、包覆層及玻璃。（Quantify the embodied carbon impacts (Global Warming Potential or GWP) of the structure, enclosure, and hardscape materials for the project. At a minimum include: asphalt, concrete, masonry, structural steel, insulation, aluminum extrusions, structural wood and composites, cladding, and glass.）

　　量化每種材料從搖籃到大門（A1 至 A3）的隱含碳排放，定義爲產品的 GWP/ 單位乘以使用的材料數量。（Quantify the cradle-to-gate (A1 through A3) embodied carbon emissions for each material, defined as the product's GWP/unit times the amount of material used.）

　　或者，使用生命週期評估（LCA）或隱含碳軟體工具的專案可以報告其工具的 A1～A3 結果。（Alternatively, projects using life-cycle assessment or embodied carbon software tools may report A1-A3 results from their tool.）

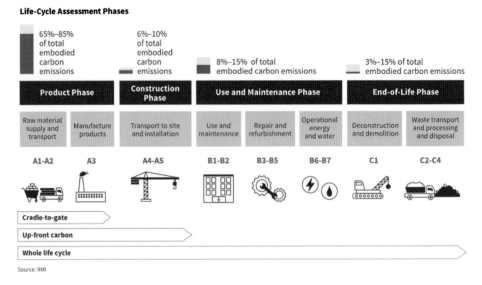

圖 8-5　生命週期評估的階段（Life-Cycle Assessment Phases）

來源：RMI

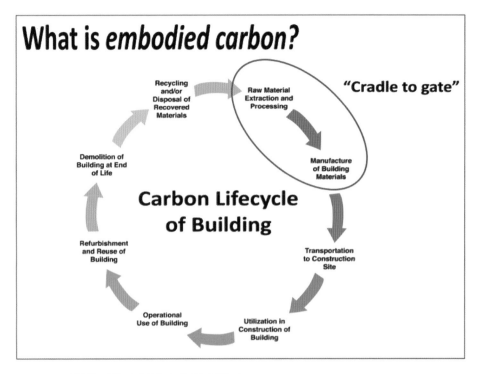

圖 8-6 建築物碳生命週期中的隱含碳（Embodied Carbon in the Carbon Lifecycle of Buildings）

來源：Northeast Sustainable Energy Association

　　此隱含碳量化結果將成為基準值，用於在 MR 得分項目：減少隱含碳中展示減排效果並獲取得分。（The results of this embodied carbon quantification become the baseline value used to demonstrate reductions and earn points in MRc: Reduce Embodied Carbon.）

高優先隱含碳來源（High-Priority Embodied Carbon Sources）

　　辨識專案中的前三大隱含碳來源，並描述專案具體策略如何被考慮以減少這些重點領域的影響。（Identify the top three sources of embodied carbon on the project and describe how project-specific strategies were considered to reduce the impacts of these hotspots.）

8.3 建築與材料再利用（Building and Materials Reuse）

以下材料被 LEED 列為可回收的類別：（The following materials are listed by LEED as recyclable categories:）

考慮回收紙類、金屬、磚塊、天花板、混凝土、塑膠、木料、玻璃、石膏牆板、地毯、隔熱材、燈具配件、表土、回填土和石頭。（**Cardboard, metal, brick, mineral fiber panel, concrete, plastic, clean wood, glass, gypsum wallboard, carpet, insulation, lighting accessories, top soil, fill dirt and rock.**）

廢棄物轉移的方法（Method of Waste Diversion）

- 回收機構：例如資源回收商或環保局（Recycling Agency）
- 轉售（Resell）
- 現場再利用（Reuse on site）
- 捐贈（Donate）

圖 8-7　回收石頭　　　　　　　　圖 8-8　放在景觀水池旁

被排除的材料（Materials to be Excluded）

挖除廢土與因清理現場所產生之廢棄物則不列入計算。（Excavated soil and land-clearing debris do not contribute to this credit.）

每日覆蓋替代材料不符合轉移處置的要求。整地所產生的廢棄物不可納入營建廢棄物回收計算。（Alternative daily cover (ADC) does not qualify as material diverted from disposal. Land-clearing debris is not considered construction, demolition, or renovation waste that can contribute to waste diversion.）

土質材料除外的材料，每個工作日結束時，放在固體廢物填埋作用面的表面，用於控制病菌、火災、氣味，防止垃圾被吹飛，以及防止人員撿拾垃圾堆裡的東西。一般來說，這些材料必須經過處理，這樣才不會在裸露的垃圾填埋場表面留下空隙。（Material other than earthen material placed on the surface of the active face of a municipal solid waste landfill at the end of each operating day to control vectors, fires, odors, blowing litter, and scavenging. Generally these materials must be processed so they do not allow gaps in the exposed landfill face.）

圖 8-9　基地表面

再利用——重複使用（Reuse）

這也是從源頭停止資源浪費的一種方式，透過重複再利用的方式來延緩或避免產品進入垃圾處理系統。（Reuse stops waste at its source because it delays or avoids that item's entry into the waste collection and disposal system.）

按照與原始應用相同或相關的功能重新使用材料，從而延長了材料的使用壽命，避免被丟棄。（The reemployment of materials in the same or a related capacity as their original application, thus extending the lifetime of materials that would otherwise be discarded.）

再利用包括回收和重新使用從既有建築或施工現場回收的材料。也稱為重新利用。（Reuse includes the **recovery** and **reemployment** of materials recovered from existing building or construction sites. Also known as **salvage**.）

「再利用二手建材」指舊房子拆除時，將可利用的材料轉賣給二手建材商，譬如木材、五金與磚頭等，由於價格便宜，加上這些二手建材已經使用一段時間，有毒揮發性氣體差不多逸散，對人體健康比較不會有影響，在國外有一定的市場，但台灣也越來越多人使用。（As for the "recycling of used building materials" refers to the removal of the old house, the material will be available sold second-hand building materials, and other such as wood, metal and brick, as cheap, plus they used building materials have been used for some time, toxic volatile gases almost escape on human health.）

「再生綠建材」係指「利用回收材料，經過再製程序，所製造之建材產品，並符合廢棄物減量（reduce）、再利用（reuse）及再循環（recycle）等原則製成之建材。」譬如回收的玻璃製成洗手台、含高比例回收再生營建廢棄物矽酸鈣板、含回收木料的木板材等等。（"Regeneration green building materials" means "the use of recycled materials, reconstituted after the procedures, manufacturing of building materials, and meet waste reduction, re use of building materials and recycling principles made it.）

再利用──建築生命週期環境衝擊評估（Reuse - Building Life-Cycle Impact Reduction）

1. 重複使用歷史建築或整建廢棄或老舊建築（Reuse historic building or renovation of abandoned or blighted building）

再利用建築物現有的結構有助於降低原始材料的使用。保持結構、外殼和內部結構元件（表面積）至少 50%：Reuse the building's existing structure can help reduce extracting raw materials. Maintain at least 50% of Structure, enclosure, and interior structure elements (by surface area)：

- 樓地板與屋頂（Structural Elements: Floors & Roof decking）
- 建築結構體與外殼（Enclosure Materials: Framing & Envelopes）
- 室內牆、門、地板材料、天花板。（Interior elements: walls, doors, floor coverings, ceiling systems.）

2. 適應性再利用（Adaptive Reuse）

設計和建造一種使建築物改變其原有用途，以適應其未來之使用功能，避免由於使用新材料而對環境造成影響。（Designing and building a structure in a way that makes it suitable for a future use different than its original use. This avoids the use of environmental impact of using new materials.）

3. 建築與建材重複使用（Building and Material Reuse）

取代垃圾掩埋場，採用回收的材料可以幫助延長材料使用壽命。再利用或者從基地外或現場回收的建築材料。（Instead of landfill, using salvaged materials can help extend the lifetime of materials. Reuse or salvage building materials from off site or on site.）

4. 重複使用、回收的、翻新的材料（**Reused, Salvaged, Refurbished** materials）

- 木樑、木柱（Beams and Posts）
- 櫥櫃和家具（Cabinetry and furniture）
- 地板（Flooring）
- 壁飾板（Paneling）
- 門與門框（Doors and Frames）
- 磚塊（Brick）
- 裝飾構件（Decorative items）

圖 8-10　裝飾構件

8.4 減少隱含碳（Reduction Embodied Carbon）

策略（Strategies）

追蹤並減少新建和翻修專案中結構、外殼及硬鋪面材料在施工過程中的隱含碳。（To track and reduce embodied carbon of major structural, enclosure, and hardscape materials from construction processes on new construction and renovation projects.）

LEED 白金級專案需達到減少 **20%** 隱含碳（A 20% reduction in embodied carbon is required for LEED Platinum projects）

全建築生命週期評估（Whole Building Life-Cycle Assessment）

進行搖籃到墓地的全建築生命週期評估（WBLCA），比較專案的結構、外殼和硬鋪面材料的結果與 MR 必要條件：評估並量化隱含碳中建立的基準值，並根據來獲得得分。（Conduct a cradle-to-grave (modules A-C, excluding operating energy and operating water-related energy) whole building life-cycle assessment (WBLCA) of the project's structure, enclosure, and hardscape materials. Compare results to the baseline developed for the MRp: Assess & Quantify Embodied Carbon and earn points.）

結果必須包括以下影響類別：（Include results for the following impact categories in the WBLCA report:）

- 全球變暖潛勢（Global warming potential, GWP），以二氧化碳當量計算（kg CO_2e）
- 平流層臭氧層耗竭（Depletion of the stratospheric ozone layer），以氯氟烴當量計算（kg CFC-11e）
- 土地和水源的酸化（Acidification of land and water sources），以氫離子當量或二氧化硫當量計算（moles H^+ or kg SO_2e）

- 富營養化（Eutrophication），以氮或磷當量計算（kg nitrogen eq or kg phosphate eq）
- 形成對流層臭氧（Formation of tropospheric ozone），以氮氧化物、臭氧當量或乙烯當量計算（kg NOx, kg O_3 eq, or kg ethene）
- 不可再生能源資源的消耗（Depletion of nonrenewable energy resources），以兆焦耳計算（MJ）或根據 TRACI 的化石燃料消耗（depletion of fossil fuels in TRACI）

　　隱含碳以全球變暖潛勢（GWP）計算，並以二氧化碳當量單位（CO_2e）表示。為了量化產品的隱含碳，使用一種名為生命週期評估（LCA）的分析來評估產品生命週期各階段的環境影響。通過環境產品聲明（EPD）披露 LCA 結果，為消費者提供有關建築產品環境影響的寶貴資訊。EPD 本質上是材料的「營養標籤」，報告包括全球變暖潛勢、酸化、富營養化、臭氧層耗竭和煙霧形成等多種生命週期影響。（Embodied carbon is calculated as global warming potential (GWP) and expressed in carbon dioxide equivalent units (CO2e). To quantify a product's embodied carbon, an analysis called life cycle assessment (LCA) is used to assess the environmental impacts associated with each stage of the product lifecycle. The disclosure of LCA results through Environmental Product Declarations (EPDs) provides valuable information to consumers about the environmental impact of building products. EPDs are essentially material "nutrition labels" that report a variety of life cycle impacts, including global warming potential, acidification, eutrophication, ozone depletion, and smog formation.）

環保產品宣告（EPD）分析：專案平均法（Environmental Product Declaration (EPD) Analysis: Project-Average Approach）

通過將專案的隱含碳與 MR 必要條件：評估並量化隱含碳中的基準值進行比較來獲取得分。專案應使用專案特定的材料數量，並識別具有特定產品或工廠類型 III EPD 的材料。（Earn points for reducing embodied carbon by comparing the project's total embodied carbon to the baseline developed in the MRp: Assess and Quantify Embodied Carbon. Projects must use project-specific material quantities and identify product-specific or facility-specific Type III EPDs for covered materials to demonstrate reductions.

Summary of **Environmental Product Declaration**	**Environmental Impacts**			
Central Concrete	Impact name	Unit	Impact per m3	Impact per cyd
Mix 340PG9Q1	Total primary energy consumption	MJ	2,491	1,906
San Jose Service Area	Concrete water use (batch)	m3	6.66E-2	5.10E-2
EF V2 Gen Use P4000 3" Line 50% SCM	Concrete water use (wash)	m3	8.56E-3	6.55E-3
	Global warming potential	kg CO2-eq	271	207
Performance Metrics	Ozone depletion	kg CFC-11-eq	5.40E-6	4.14E-6
	Acidification	kg SO2-eq	2.26	1.73
28-day compressive strength — 4,000 psi	Eutrophication	kg N-eq	1.31E-1	1.00E-1
Slump — 4.0 in	Photochemical ozone creation	kg O3-eq	46.6	35.7

A sample EPD for a concrete mix design by Central Concrete Supply Co.
Credit: Central Concrete Supply

圖 8-11 環境產品聲明與產品類別規則（Environmental Product Declarations And Product Category Rules）
來源：US.Federal Highway Administration

環保產品宣告（EPD）分析：按材料類別（Environmental Product Declaration (EPD) Analysis by Materials）

透過展示針對選定材料的隱含碳減少，相較於行業基準值來獲得得分，並根據附表計算得分。專案可以根據產品類別的加權平均隱含碳強度來計算減少量。（Earn points according to Table by demonstrating that structural, enclosure, and hardscape materials for targeted materials have lower-embodied carbon impacts than industry benchmarks.

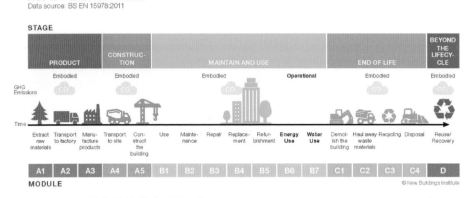

圖 8-12　隱含碳的生命週期（carbon throughout the life cycle of a building）

來源：New Buildings Institute

追蹤施工活動中的碳排放（Track Carbon Emissions from Construction Activities）

根據追蹤施工活動期間的碳排放來獲取得分。（Earn points for tracking carbon emissions during construction activities）

表 1：選項 1～3 的隱含碳減少得分（Table 1. Points for embodied carbon reductions in Options 1-3）

減少隱含碳百分比（Reduction in GWP）	選項 1：全建築生命週期評估（Option 1: WBLCA）	選項 2：EPD 專案平均法（Option 2: Project-Average EPD）	選項 3：按材料類別分析（Option 3: EPD by Materials）
符合基準值或行業平均（Meet Baseline or Industry Average）	2 分（2 points）	1 分（1 points）	3 個材料類別 1 分（3 material categories for 1 point）或 5 個材料類別 2 分（or 5+ material categories for 2 points）
減少 10% GWP（10% Reduction in GWP）	3 分（3 points）	-	-
減少 20% GWP（20% Reduction in GWP）	4 分（4 points）	2 分（2 points）	-
減少 30% GWP（30% Reduction in GWP）	5 分（5 points）	-	-
減少 40%+ GWP（40%+ Reduction in GWP）	6 分（6 points）	3 分（3 points）	-

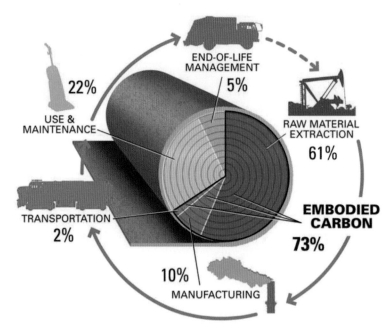

圖 8-13　尼龍地毯的碳足跡（carbon footprint of Nylon Carpet）

來源：Interface Global. Illustration: Peter Harris

　　建築營運和與建設相關的活動約占全球人類溫室氣體（GHG）排放的 39%。其中超過四分之一是隱含碳排放，這些排放與建材生產、建設活動、運營以及建築壽命終結階段相關。此外，用於製造製冷劑的含氟氣體（F-gas）也占所有溫室氣體排放的 2%。隱含碳排放比運營排放更難以察覺，並且長期以來基本被忽視，但隱含碳也是導致環境退化的重要排放來源。主管機關、製造商、建築師、工程師和承包商都應在減輕這些排放影響方面發揮作用。（Building operations and construction-related activities are responsible for approximately 39% of humanity's global greenhouse gas (GHG) emissions. More than a fourth of those are embodied carbon emissions, those associated with the production of building materials, construction activities, operations, and end of life. In addition, fluorinated gases (F-gas), used to make refrigerants, are responsible for another 2% of all GHG emissions. Embodied emissions are less discernible than operational emissions and have been largely ignored, yet embodied carbon is an equally important emission source causing environmental degradation. Regulators, manufacturers, architects, engineers, and contractors all have a role to play in mitigating the impacts of these emissions.）

8.5 低排放材料（Low-Emitting Materials）

對於新建築（建築物或建築物的部分），進行生命週期評估，比較建物的結構和外殼與基準建築的值，顯現其最少減低 10%，至少有三個下面列出的六個影響類別。（For new construction (buildings or portions of buildings), conduct a life-cycle assessment of the project's structure and enclosure that demonstrates a minimum of 10% reduction, compared with a baseline building, in at least three of the six impact categories listed below.）

- 全球暖化潛勢（溫室氣體）（Global warming potential (greenhouse gases)）（單位＝CO_2e）也稱爲 GWP。
- 平流臭氧層破壞（Depletion of the stratospheric ozone layer）（單位＝kg CFC-11）也稱爲 ODP。
  1. ODP 值，是「臭氧破壞潛力」的英文縮寫，它表示物質分子分解臭氧的能力。ODP 的數值以 R11 爲基準，規定 R11 的 ODP 值爲 1，其他物質的 ODP 值按其相對 R11 損耗臭氧能力來計算大小。（ODP (Ozone Depletion Potential): This value represents a substance's ability to break down ozone molecules. It is measured relative to R11, where R11 has an ODP value of 1. Other substances are compared based on their ozone depletion potential relative to R11.）

2. GWP 值，是「全球變暖潛力」的英文縮寫，是氣體溫室效應的評價標準。（GWP (Global Warming Potential): This value assesses the potential of a gas to contribute to the greenhouse effect and global warming. It provides a standard to compare the impact of different gases.）

- 土壤與水資源酸化（Acid ification of land and water sources）（單位 = moles H^+ or kg SO_2）

- 優養化（Eutrophication）（單位 = kg nitrogen 或 kg phosphate）

- 對流臭氧層形成（Formation of tropospheric ozone）（單位 = kg NOx 或 kg ethene）

- 非再生能源資源耗盡（Depletion of nonrenewable energy resources）（單位 = MJ）

8.6 建材披露與優化（Building Product Disclosure and Optimization）

建材聲明（Building Product Disclosure and Optimization）

考慮生命週期評估（Considering Life-Cycle Assessment (LCA)）

生命週期評估是一種從搖籃到墳墓（Cradle to Grave）的衝擊評估方法，也就是從物品的製造、使用、回收到最後廢棄階段，估算整體能源和原料的取得、運輸、產品使用、消耗等；近來也有將評估範疇更延伸發展至廢棄品再利用的趨勢，也就是搖籃到搖籃的評估。（Life Cycle Assessment (LCA): is the investigation and valuation of environment impact of building the throughout its life span）

圖 8-14　生命週期評估

環保材料證書（Environmental Product Declarations (EPD)）

1. 漂綠（Greenwashing）

　　泛指將大量的金錢和時間花費在廣告自身產品是綠色環保的，實際上對於環境保護所花的時間與精力卻比廣告還少。（The term greenwashing is generally used when significantly more money or time has been spent advertising being green, rather than spending resources on environmentally sound practices.）

www.lorenafdezblog.com

圖 8-15　漂綠產品上面常有「Nature」、「Green」等字樣，誤導消費者。

產品透明度（Product Transparency）

1. 建材聲明（Disclosure）

這是一個關於產品的影響成份表，在以下方面：（This is about being transparent about a product's impact on the following aspect:）

- 環境衝擊（Environmental impacts）
- 健康衝擊（Health impacts）
- 原物料開採（Sourcing of raw materials）

圖 8-16　EPD 聲明有機食材之成分標示

- 材料成分（Material Ingredients）

2. 環保材料證書（產品環境聲明）Environmental Product Declarations (EPD)

基於產品生命週之特性，依據 ISO 標準「產品環境標誌與宣告」，提供消費者量化且可比較之環境績效結果，至少須提供搖籃到大門的範圍。（EPD is a standardized way of quantifying the environmental impact of a product, base on the environmental requirements of ISO 14021–1999, ISO 14025–2006 and EN 15804, or ISO 21930–2007, and have at least a **cradle to gate** scope.）

產品第三類環境宣告（EPDs, formally called "Type III Environmental Product Declarations" by ISO）

- 工業界（一般）環境宣告（Industry-wide (generic) EPD）
- 產品環境宣告（Product-specific Type III EPD）

選用最少 20 種永久安裝性材料，其材料來自最少 5 家不同的廠商。

（Use at least 20 different permanently installed products from at least five different manufacturers.）

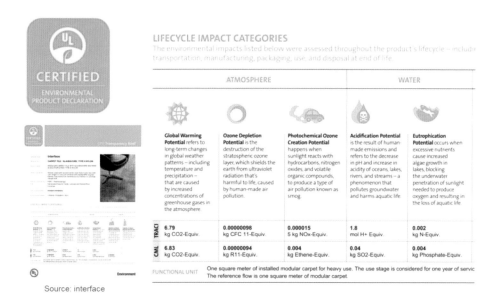

Source: interface

圖 8-17　EPD 環保材料聲明之範例

MATERIAL CONTENT
Material content measured to 1%.

COMPONENT	MATERIAL	AVAILABILITY	MASS%	ORIGIN
Face Cloth/Yarn	Nylon 6 Post Industrial & Post Consumer Recycled	Recycled material, abundant	17%	IT
Tufting Primary	Polyester	Fossil resource, limited	2%	US
Latex	Ethylene vinyl acetate	Fossil resource, limited	5%	US
Filler	CaCO3	Mineral resource, non renewable, abundant	14%	US
Fiberglass	Silica	Mineral resource, non renewable, abundant	1%	US
GlasBac®RE Backing	Post Consumer recycled carpet tile	Recycled material, abundant	26%	US
GlasBac®RE Backing	Post Industrial recycled vinyl	Recycled material, abundant	35%	US

ENERGY

RENEWABLE ENERGY	2 %	2.32 MJ
NON-RENEWABLE ENERGY	98 %	104.03 MJ

RECYCLING OR REUSE

Product should be recycled through Interface's ReEntry® 2.0
process by contacting Interface ReEntry® 2.0 at
888-733-6873 (US) or 866-398-3191 (Canada).

ADDITIONAL ENVIRONMENTAL INFORMATION

PRE-CONSUMER RECYCLED CONTENT	52 %
POST-CONSUMER RECYCLED CONTENT	26 %
VOC EMISSIONS	CRI GLP Certified
WATER CONSUMPTION	0.02 m3

CERTIFICATIONS

Source: interface

圖 8-18　材料成分標示之範例

原物料開採報告（Raw Material Source and Extraction Reporting）

1. 製造商的開採報告（Manufacturers' Self-Declared Reports）

此產品的 1/2 金額可貢獻於取分（valued as one half (1/2) of a product for credit achievement）

2. 企業永續報告書（Corporate Sustainability Reports (CSRs)）

第三方驗證的企業永續報告書內容完整記載產品與其供應鏈因開採與製造過程所造成的環境影響，符合的企業永續報告書規範如下（Third-party verified CSRs help to identify products/manufacturer's that have been verified to be **extracted or sourced in a responsible manner**.）：

* 全球報告倡議組織（GRI）可持續發展報告（Global Reporting Initiative (GRI) Sustainability Report）

* 經濟合作與發展組織（OECD）跨國公司指南（Organization for

Economic Co-operation and Development (OECD)Guidelines for Multinational Enterprises）

- 聯合國全球契約：溝通的過程（U.N. Global Compact: Communication of Progress）
- ISO 26000：2010 社會責任指南（ISO 26000: 2010 Guidance on Social Responsibility）
- USGBC 同意之計畫：其他 USGBC 批准符合 CSR 的標準程序。（USGBC approved program: Other USGBC approved programs meeting the CSR criteria.）

領導性開採手法（LEADERSHIP EXTRACTION PRACTICES）

1. 生物基材料（Bio-Based Materials）

產品應符合永續農業網的永續農業規範。（Bio-based products must meet the **Sustaina ble Agriculture Network's Sustaina ble Agriculture Standard**.）

生物基原物料需採用 ASTM D6866 測試，並且爲合法取得。皮革產品（例：牛皮、其他動物皮料）不納入計算（Bio-based raw materials must be tested and be legally harvested. Exclude hide products, such as leather and other animal skin material.）

2. 快速可再生材料（Rapidly Renewable Materials）

快速可再生材料是指十年之內可以長成並採收利用的材料（大多爲植物性產品，例如竹材、軟木、棉麻等製品）。（Building materials that are quickly grown or raised and can be harvested in a sustainable fashion. For LEED, rapidly renewable materials take **10 years** or less to grow or raise.）

圖 8-19　可快速再生材料之示意

3. 木材產品（Wood Product）

　　木材產品需爲森林管理委員會（FSC）認證，或其他 USGBC 認可的規範。（Wood products must be certified by the Forest Stewardship Council or USGBC-approved equivalent.）

- 認證木料指具有永續森林管理委員會（FSC）認證之木材產品，包括結構材和一般尺寸木角材、木地板材、地板底層材、木門與表面裝修材等。（**Certified Wood**: Wood that has been issued a certificate from an independent organization with developed Certified Wood standards of good forest management.）

- 產銷監管鏈用於追蹤產品的流程，從開採或取得到其最終用途，包括加工、轉化、製造及分銷這幾個連續的階段。（Chain-of-custody (COC): a tracking procedure for a product from the point of harvest or extraction to its end use, including all successive stages of processing, transformation, manufacturing and distribution.）

圖 8-20　FSC 木材具有永續循環的概念

- 森林管理委員會（Forest Stewardship Council (FSC)）：為確保林木資源健康發展，並且為森林管理〔Forest Management（FM）〕及產銷監管鏈〔Chain-of-Custody（CoC）〕建立國際的標準。透過 FSC-FM 森林管理以及 FSC-CoC 產銷監管鏈，可以保證您使用的產品是來自於栽培的森林，做到「可追溯性」與「森林管理」。

4. 消費後回收成分產品（Post-Consumer Recycled Content）

　　其定義為經住家、商業、工業或機構等設施使用後之廢料，該廢料必須是使用者使用過，且已無法再回收作為原始功能使用。（例：玻璃瓶飲料經過消費者使用完畢後被丟棄，廠商把玻璃瓶打碎回收重複利用，製成其他的材料或成品）。（The percentage of material in a product that was consumer waste. The recycled material was generated by household, commercial, industrial, or Post-Consumer institutional end users and can no longer be used for its intended purpose.）

- 鋁罐（Aluminum cans）

- 水瓶（Water bottles）

- 塑膠瓶（Plastic bottles）

- 報紙（Newspapers）

- 建築殘骸（Construction debris）

- 舊家具／櫥櫃（Old furniture/Cabinetry）

- 景觀綠化廢棄物（Landscaping wastes）

相關規範（Reference Standards: ISO 14021）

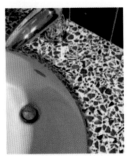

圖 8-21　使用回收的玻璃瓶再製造成流手台的材料

5. 消費前回收成分產品（Pre-Consumer Recycled Content）

其定義為製造過程中所產生之廢物經回收後再利用，且該材料不得經過重新加工或處理後，再回到原來之生產線。（例：家具生產過程所產生的木屑，被回收製成木板）。（The percentage of material in a product that was recycled from manufacturing waste. Pre-consumer content was formerly known Pre-Consumer as postindustrial content (by-product).）

- 鋸屑（Sawdust）
- 木片（Wood chips）
- 樹皮（Tree bark）
- 氾濫的雜誌（Magazine overruns）
- 飛灰（Fly-ash）
- 刨花（planer shavings）
- 核桃殼（walnut shells）
- 修整的材料（trimmed materials）
- 過時的存貨（Obsolete inventories）

圖 8-22　木屑與纖維板
（密集板）

本地建材（Regional/Locally Produced Material）

　　若材料於距離基地 100 英哩（160 公里）範圍內開採、運輸、取得、加工製造以及購買，則其材料購買金額的 200% 可貢獻於取分；例如：買 $100 本地建材，可以當作 $200 的得分使用。（Building materials sourced (extracted, manufactured, purchased) within 100 miles (160 km) of the project site are valued at 200% of their base contributing cost.）

圖 8-23　例如位於美國中北部之基地，其材料需從 100 英哩以內範圍開採、製造、使用。

建材成分聲明報告（Material Ingredient Disclosure）

1. 健康產品宣告（Health Product Declaration (HPDs)）

完整公開的健康產品宣告，明確記載根據健康產品宣告公開規範（Health Product Declaration Open Standard）所定義的危害物質。（Health Product Declaration (HPDs) provide a full disclosure of the potential chemicals of concern in products by comparing product ingredients to a wide variety of "hazard" lists published by government authorities and scientific associations.）

2. 搖籃到搖籃（Cradle to Cradle (C2C)）

搖籃到搖籃強調一個無浪費的產品，透過從搖籃到搖籃循環材料的封閉系統永遠流傳。無論是產品和副產品都具有價值。從搖籃到搖籃的設計考慮了產品能被重新再利用的機會。（Through the cradle-to-cradle cycle the materials are perpetually circulated in closed loops – essentially a waste-free product. Both the product and any byproducts have value. Cradle-to-cradle design considers opportunities for the reuse of the product.）

- 列出所有不含有等級 1 危害物質的化學成分（GreenScreen v1.2 Benchmark）
- 國際通用的替代性作法（REACH Optimization(international project)）
- USGBC 認可的方案（USGBC approved program）

從搖籃到搖籃（Cradle to Cradle(C2C)）

從搖籃到墳墓 / 搖籃到大門（Cradle to grave/Cradle to gate）

圖 8-24　永續環境讓地球生生不息

圖 8-25　上：搖籃到搖籃商標；下：搖籃至搖籃經典著作（作者：威廉麥唐諾）

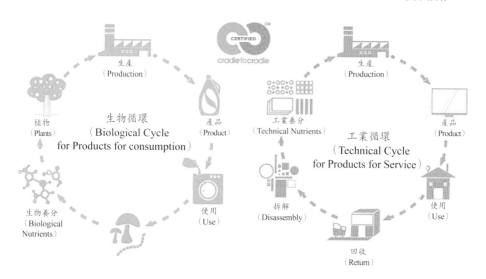

圖 8-26　搖籃到搖籃開發模式

名詞小教室

1. 二手回收建材（Salvaged material/ Reused material）

　　二手建材乃是指工地 / 現場回收後再利用，原來的功能還在。（Salvaged (reused) materials come from construction sites or existing buildings and are reused in the same or different capacity.）

2. 翻新二手建材（Refurbished material）

　　翻新的手法意旨翻新、修復等改善其外觀、性能、品質，使其功能與價值等同於全新品。（Refurbishing includes renovating, repairing, restoring, or generally improving the appearance, performance, quality, functionally, or value of a product.）

3. 再生建材 / 回收成分建材（Recycled (Content) material）

　　再生建材的成分通常使用回收的材料製作而成，這些被回收的材料會與原始的功能不同，例如消費前、消費後回收成分建材。(Recycled content contains materials that have been recycled. Usually the recycled materials will not be perform the original function. (ie. Post- Consumer & Previous Consumer Materials)）

永續性採購計畫（Sustainable Purchasing Program）

(1) 施工材料採購計畫（Develop a construction purchasing policy）

　　並非所有的團隊成員都能理解你的設計目標。採購政策可以幫助確保成員不會有疑問。（Not all team members can understand your goal. Put together a purchasing policy can help make sure there's no question about it.）

(2) 使用綠色環保建材（Specify green materials）

　　建築師和設計師應指定的綠色環保材料，如快速再生材料、地方材料、二手回收的材料以及材料的再生含量。（Architect and designer should specify green materials, such as rapidly renewable materials, regional materials, salvaged materials, and materials with recycled content.）

備註：其他專有名詞：翻新，重複使用（NOTE: Other key terms to note: refurbished, reused）

(3) 制定綠色採購計畫（Develop a sustainable purchasing policy）

提供綠色持續消耗品和耐用品一個清單，以確保每個人都知道目標和監控遵守，並確保該政策是有效的。（Simply provide a checklist of green ongoing consumable and durable goods to ensure everyone understands the goal and monitor the compliances to ensure the policy is effective.）

- 使用環保省電電器設備（Specify green electronic equipment）選擇那些使用較少能源的電腦及家用電器。（Choose computers and appliances that uses less energy.）例如具有 Energy Star 標章或省電標章的設備。

- 使用低含汞燈泡（Specify low elemental mercury light bulb）購買很低或沒有汞的燈泡。（Purchase bulbs with very little or no mercury.）

- 採購認證的食材（Certified Food Products）如公平貿易標籤的產品、美國農業部有機食品、雨林聯盟認證的食材。（Such as Fair trade Labeled products, USDA organic food, Rainforest Alliance Certification）

2. 固體廢棄物管理的階層（Solid Waste Management Hierarchy）

源頭減量 （Source Reduction）	適用性再利用 （Adaptive Reuse）	再利用的材料 （Reused Materials）	二手回收材料 （Salvaged Materials）	翻新的材料 （Refurbished Materials）
回收 （Recycle）	消費前回收 成分產品 （Pre-consumer ReCycled Content）	消費後回收 成分產品 （Post-consumer Cycled Content）	本地材料 （Regional Materials）	快速可再生材料 （Rapidly Renew- able Materials）
木材認證（FSC） （Certified wood (FSC)）	混合廢棄物 （Commingled Waste）	廢棄物轉移 （Waste Diversion）	廢棄物轉換能量 （Waste to Energy）	垃圾掩埋場 （Landfill）
搖籃到搖籃 （Cradle-to-Cradle）	搖籃到墳墓 （Cradle-to-Grave）	搖籃到大門 （Cradle-to-Cate）	生命週期評估 （Life-cycle assess- ment (LCA)）	生命週期成本 （Life-cycle cost- ing (LCC)）

8.7 施工與拆除廢棄物轉移（Construction and Demolition Waste Diversion）

策略（Strategies）

1. 鼓勵回收（Encourage recycling）

- 規劃資源回收收集與存放區域，並且確認清運貨車與建築人員可方便進出與使用。（Provide a suitable and convenient storage facility for occupants to store recyclable materials and the haulers to periodically collect the recyclable materials.）
- 回收項目必須包括：紙類、玻璃、塑膠與金屬等。（Must include: paper, corrugated cardboard, glass, plastics and metals.）

圖 8-27　資源分類回收區　　　　圖 8-28　鋁罐壓扁器

- 以下至少兩項廢棄物：電池、含汞燈泡與電子產品。必須規劃安全的收集、存放與丟棄方式。（Take appropriate measures for the safe collection, storage, and disposal of two of the following: batteries, mercury-containing lamps, and electronic waste.）

- 回收計畫內容含耐久產品，例如：家具、電器產品，以及家電類。（Recycling can also cover durable goods, such as furniture, electronics, and appliances.）

2. 基礎設施規劃考量（Infrastructure considerations）

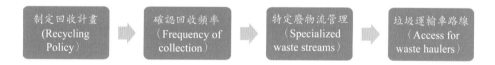

| 制定回收計畫
(Recycling Policy) | ⇒ | 確認回收頻率
（Frequency of collection） | ⇒ | 特定廢物流管理
（Specialized waste streams） | ⇒ | 垃圾運輸車路線
（Access for waste haulers） |

營建與拆除廢棄物管理計畫（Construction and Demolition Waste Mgmt Planning）

- 營建與拆除廢棄物約占美國總固體廢物流的 40%，歐盟地區則占了約 25%，屬於相當多的比例。（Construction and demolition waste constitutes about 40% of the total solid waste stream in the United States and about 25% of the total waste stream in the European Union.）

http://landfill-site.com

圖 8-29　施工廢棄物管理

- 營建廢棄物管理計畫（Construction Waste Management Plan）建立營建廢棄物回收目標，明確列出至少五種計畫可回收材料（包含結構與非結構）。（Prepare a construction waste management plan in order to divert construction waste from landfill. Establish waste diversion goals for the project by identifying at least 5 materials (both structural and nonstructural) targeted for diversion.）

- 導入施工廢棄物管理計畫，回收或再利用至少 50% 或 75% 之營建工程廢棄物。（Develop and implement a construction waste management plan target 50% or 75% recycling goal.）

- 計算方式可以重量或體積來進行，但在整體計算過程中必須是統一的單位。（Calculations can be done by **weight** or **volume**, but must be consistent throughout.）

- 回收材料可於工地分類，或混合統一由回收廠商自行分類。（Recycled materials can be **sorted on-site** or **commingled**.）

混合回收（Definition: Commingled recycling）建築廢棄物流在工程現場中混合合併運走，但不包含不可回收的項目。（when all recyclables for collection are mixed but kept separate from other waste.）

圖 8-30 混合回收　　　　　　　　圖 8-31 分類回收

名詞解釋（Glossary）

- 產銷監管鏈（Chain-of-custody）

產銷監管鏈用於跟蹤產品的程序，從收穫或萃取到其最終用途，包括加工、轉化、製造及分銷這幾個連續的階段。（A tracking procedure for documenting the status of a product from the point of harvest or extraction to the ultimate consumer end use, including all successive stages of processing, transformation, manufacturing, and distribution.）

- 混合回收（Comingling Recycling）

回收材料的過程中，可以允許消費者放置各種材料（如紙張、紙板、塑料和金屬）在一個容器，它是獨立於廢棄物的。在它們被收集並帶到資源回收廠前，這些回收材料都沒有被分類。（A process of recycling materials that allows consumers to dispose of various materials (such as paper, cardboard, plastic, and metal) in one container that is separate from waste. The recyclable materials are not sorted until they are collected and brought to a sorting facility.）

- 複合木 / 合板（Composite Wood）

  複合木由幾種材料製成，農業纖維製品是從農業纖維製成的產品。這些材料包括刨花板、中密度纖維板（MDF）、膠合板、刨花板（OSB）、小麥板、硬紙板、面板基板和門芯。（Composite wood is made from several materials, and agrifiber products are products made from agricultural fiber. For this credit, these materials comprise particleboard, medium-density fiberboard (MDF), plywood, oriented-strand board (OSB), wheat board, strawboard, panel substrates, and door cores.）

- 家具、固定裝置和設備（Furniture, Fixtures, and Equipment）

  包括所有項目，不屬於基礎建築構件，例如燈、計算機和電子產品、書桌、椅子、桌子。（Include all items that are not base building elements, such as lamps, computers and electronics, desks, chairs, and tables.）

- 源頭減少（Source Reduction）

  減少不必要的材料進入建築物。包括採購產品、選擇更少的包裝和可持續發展的設計。（Method of reducing the amount of unnecessary material brought into a building. Examples include purchasing products with less packaging and sustainable design.）

- 永續性採購政策（Sustainable Purchasing Policy）

  優先選擇在整個生命週期小到對環境和對社會沒有負面影響的產品，並優先選擇產品公司也幾乎沒有任何負面的環境和社會影響。可持續採購政策讓該單位員工營運時，持續產生影響。（A policy that gives preference to products that have little to no negative impacts on the environment and society throughout its life cycle, and also gives preference to those products that are supplied by companies whom also have little to no negative environmental and social impacts. The sustainable purchasing policy commits the organization to an overarching course of action, which empowers staff working at the operations level.）

- 永續性採購程序（Sustainable Purchasing Program）

開發、採用和實施的採購策略，最終購買那些整個生命週期小到對環境和社會沒有負面影響的產品，或其產品公司也幾乎沒有任何負面的對環境和社會影響。該方案策略是採購一致並支持企業的永續性採購政策。

（The development, adoption, and implementation of a procurement strategy, which culminates in the purchase of products that have little to no negative impacts on the environment and society through its life cycle or that are supplied by companies whom also have little to no negative environmental and social impacts. The program is an operational working strategy aligned and in support of an organization's sustainable purchasing policy.）

- 廢棄物轉移（Waste Diversion）

一個處置廢棄物管理的活動，勝於焚燒或填埋等。例如再利用和再循環。（A management activity that disposes of waste other than though incineration or landfilling. Examples are reuse and recycling.）

- 減少廢棄物（Waste Reduction）

包括透過再利用和再循環的源頭減少及廢物轉移。（Includes source reduction and waste diversion through reuse or recycling.）

- 廢棄物流向（Waste Stream）

來自建築物的整體廢棄物流向掩埋場、焚燒處、回收設施或其他處理場所。（The overall flow of wastes from the building to the landfill, incinerator, recycling facility, or other disposal site.）

題目（Practice Questions）

1. 建築物的材料是從回收的飲料瓶製造的，此成份稱爲？（A building material that is made from recycled soda bottles contains ？）

   a. 工業後回收成分產品（Post-industrial recycled content）

   b. 消費後回收成分產品（Post-consumer recycled content）

   c. 消費前回收成分產品（Pre-consumer recycled content）

   d. 預製前回收成分產品（Pre-fabricated recycled content）

   b. 消費後回收的內容來源於消費後的廢棄物，與消費前回收的內容是來自製造過程不一樣（Postconsumer recycled content comes from consumer waste, contrasted with pre-consumer recycled content, which comes from manufacturing processes.）

2. 以下哪項是源頭控制減少廢棄物策略的例子？（Which of the following is an example of a source control waste reduction strategy?）

   a. 預購材料切割成大小（Pre-ordering materials cut to size）

   b. 使用混合回收（Using commingled recycling）

   c. 重複使用現場的二手材料（Reusing salvaged materials onsite）

   d. 現場安裝回收箱（Installing recycling bins onsite）

   a. 來源減少最好的方式是減少浪費，源頭減少的方式是從預製的材料切割成一定的尺寸，並選擇產生較少的工地廢棄物的模組化結構。（Source reduction is the first and best way to minimize waste. Source reduction starts at the source-such as pre-ordering materials cut to size and choosing modular construction, which generates less onsite waste.）

3. 成功的廢物管理政策的第一步是？（What is the first step in a successful waste management policy?）

a. 回收所有可能的材料（Recycle all possible construction materials）

b. 再利用現場材料（Reuse existing materials）

c. 減少廢棄物總量（Reduce the total quantity of waste）

d. 指定可回收的材料（Specify recyclable materials）

e. 確定產品的能耗（Determine the embodied energy of the product）

c. 減少廢物是首先需考慮的廢物管理政策，其次是尋求機會以便再次使用。然後回收應考慮廢棄物其不能被消除或改變用途。（Waste reduction should be the first consideration of a waste management policy, followed by exploring opportunities for reuse. Recycling should then be considered for waste streams that can't be eliminated or repurposed.）

4. 製造商有一張地毯，其中包括一個環境產品聲明（EPD）和健康產品聲明（HPD）。製造商還公佈了企業可持續發展報告（CSR）。這些文件有助於？（A manufacturer has a carpet tile that includes an environmental product declaration (EPD) and health product declaration (HPD). The manufacturer has also published a corporate sustainability report (CSR). Which of the following do these documents aid?）

a. 改善社區（Improving the community）

b. 降低產品成本（Reducing product costs）

c. 公開透明資訊（Transparency）

d. 材料優化設計（Material design optimization）

c. 透明度在材料得分中包含：（Transparency in the materials credits comes from publishing:）

- EPD，它包括一個產品的壽命週期帶來的影響。（EPDs, which include the life-cycle impacts of a product.）

- CSR 以負責任的態度幫助確定已證實被提取或態度來源的產品／生

產廠家。（CSRs help to identify products/manufacturers that have been verified to be extracted or sourced in a responsible manner.）

- 報告產品內容和建築相關的健康資訊。（HPD reports product content and associated health information for building.）

5. 持續進行的消耗品在以下哪一個期間消耗？（Ongoing consumables are consumed during which of the following?）

a. 只有設計（Design only）

b. 只有建設（Construction only）

c. 只有進駐期間（Occupancy only）

d. 包含建設與經營（Both construction and operation）

e. 建築生命的結束和建設（Construction and at the building's end of life）

c. 持續進行消耗品是每單位定期使用與在業務過程中一直消耗的產品（例如紙、電池和肥皂）。（Ongoing consumables are goods with a low cost per unit that are regularly used and replaced in the course of business (for example, paper, batteries, and soap).）

6. 下列哪些是減輕其整個生命週期之內部空間的整體環境結果的有效方式？（Which of the following are effective ways to lessen the overall environmental consequences of an interior space over its lifetime?）

a. 選擇基地距離公共交通點位於 1 英里（1600 公尺）（Selecting a site located 1 mile (1,600 meters) from public transportation）

b. 安裝可拆卸的內部非結構牆（Installing demountable interior non-structural walls）

c. 安裝 1.6 GPF（6 LPF）廁所（Installing 1.6 gpf (6 lpf) toilets）

d. 選擇具有發佈環境產品聲明的產品（EPD）（Choosing products that have a published Environmental Product Declaration (EPD)）

e. 選擇翻新的家具（**Selecting refurbished furniture**）

b. 這種策略是設計的一部分，減少隨著時間需求的新型建築空間的靈活性。（This strategy is part of designing for flexibility which reduces the demand for new building materials over time.）

e. 重複使用的材料或家具減少了對新材料的需求，從而保持原始資源。（Reusing materials or furnishings reduces the demand for new materials and thus preserves virgin resources.）

7. LEED 所認定「可快速再生材料」為幾年內生長？（How many years can an agricultural product grow or be raised to be considered as rapidly renewable by LEED?）

a. 5

b. 10

c. 15

d. 20

b. 快速的可再生資源，LEED 定義種植／收穫週期在 10 年或 10 年以下，可比傳統材料得到更快的補充。（Rapidly renewable resources, defined by LEED as having a planting/harvest cycle of 10 years or less, are replenished more quickly than conventional materials.）

第九章　室內環境品質（Indoor Environment Quality）

學習目標（Learning Objectives）

基地選擇對於綠建築設計規劃的重要性，對於建築物生命週期的影響，可由下列幾個面向來討論：（The importance of site selection in green building design and its impact on the life cycle of a building can be discussed from several perspectives:）

- 室內空氣品質（Indoor Air Quality）
- 熱舒適（Thermal Comfort）
- 照明（Lighting）
- 聲音（Acoustics）
- 提升室內環境品質大幅提高生產力。包含減少員工病假時數、減少過敏與氣喘、提高工作表現。（Improving indoor air quality can increase productivity. The effect includes reducing employee sick hours, reducing allergies and asthma, and increasing worker's performance）

Triple Bottom Line

圖 9-1　三重底線（三大原則）

- 人們大部分時間都在室內活動，提供適當但不過量的新鮮空氣，可以在節能與健康找到平衡點。（People spend most of their time:

indoors. Proving ample, but not excessive ventilation strikes the right balance between energy use and human health.）

- 好的室內環境是理想的生活與聚會地點。健康的室內環境也能減少疾病的發生，改善社區成員的健康。（High-quality indoor air are desirable place to live, and ideal gathering place. Improvement to indoor quality can reduce of diseases and improve people's health.）

9.1 施工管理計畫（Construction Management Plan）

策略（Strategies）

- 維護建造期間的空氣品質（Protect air quality during construction）
  保護工人和設備免受灰塵和濕氣破壞。（Protect worker and equipment from dusts and moisture damage.）
- 進行大量換氣（Conduct a flush out）
  提供大量的空氣，除去來自建造活動的汙染物。（Deliver large volume air to remove contaminants from construction activity.）
- 使用整合害蟲管理（Use integrated pest management）
  盡量減少人體暴露於有害生物防治化學品。（Minimizing human exposure to pest-control chemicals.）
- 執行綠色清潔計畫（Employ a green cleaning program）

圖 9-2　清潔時選用綠色清潔產品

保障施工期間的空氣品質（Protect Air Quality During Construction）

- 在施工期間與使用者進駐前，擬定並進行以下的室內空氣品質管理計畫。（Develop and implement an IAQ management plan for the construction and preoccupancy phases.）
- 室內空氣品質管理計畫要符合美國國家鈑金與空調包商協會（SMACNA）的室內空氣品質管理計畫指導方針之建議。（The IAQ plan meets or exceeds the Sheet Metal and Air Conditioning National Contractors Association (SMACNA) IAQ Guidelines For Occupied Buildings Under Construction.）

圖 9-3　美國國家鈑金與空調包商協會

- 妥善保護或存放基地中之具吸收性的建材，例如板材與隔熱棉等，以避免其受潮或汙染。（Protect stored on-site and installed absorptive materials from moisture damage.）

圖 9-4　板材與隔熱棉

- 假如空調系統已被固定安裝，並已在施工期間開動啓用，於每個回風口更換空氣濾網，該濾網並應符合 ASHRAE 52.2-1999 建議之最低效率值（MERV）8。（If permanently installed air handlers are used during construction, filtration media with a minimum efficiency reporting value (MERV) of 8 must be used.）

進行吹洗（空氣換氣）（Conduct a Flush Out）

在使用者進駐前，擬定並進行以下的室內空氣品質管理計畫：（Before occupancy, develop and implement the following Indoor Air Quality Management Plan:）

1. 吹洗（Option 1: Flush-Out）

- 在施工完成後，使用者進駐前以及所有內裝完成後，進行建築物的換氣。（After construction ends, prior to occupancy and with all interior finishes installed, install new filtration media and , perform a building flush-out by supplying outdoor air to indoor.）

2. 空氣品質檢測（Option 2: Air Testing）

- 在施工完成後，使用者進駐前以及所有內裝完成後，進行建築物室內空氣品質檢測。（Conduct baseline IAQ testing after construction ends and prior to occupancy.）

採用低逸散性建材（Specify Low-Emitting Material）

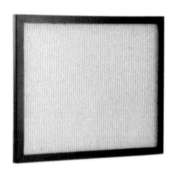

圖 9-5　低逸散性之建材

　　所有使用於建築物內部裝修的黏著劑與填縫劑應符合以下標準（All adhesives and sealants used on the interior of the building must comply with the following requirement）：

- 黏著劑、填縫劑與填縫劑底塗：加州南岸空氣品質管理局（SCAQMD）規範第 1168 條。（Adhesives and Sealants and Sealant Primers：South Coast Air Quality Management District (SCAQMD) Rule #1168.）
- 噴霧黏著劑：美國綠色標章（Green Seal）標準之商用黏著劑規範 GS-36（Aerosol Adhesives: Green Seal Standard GS 36）

圖 9-6　黏著劑標章　　圖 9-7　黏著劑與塗料

- 所有使用於建築物內部裝修的塗料應符合綠色標章（Green Seal）的標準。（Paints and coatings used on the interior of the building (i.e., inside of the weatherproofing system and applied onsite) must comply with Green Seal the following criteria.）

圖 9-8　各式油漆

- 所有安裝於建築物室內的地毯應符合美國地毯協會（CRI）的 Green Label Plus Program 產品測試需求。（All flooring must comply with the requirement of Green Label Plus Program of Carpet and Rug Institute (CRI).）

圖 9-9　地毯　　　　　　　　　圖 9-10　地毯標章

- 混凝土、木材、竹製與軟木地板之表面材必須符合南岸空氣品質管理局 (SCAQMD) Rule 1113。（Concrete, wood, bamboo and cork floor finishes must meet the requirements of South Coast Air Quality Management District (SCAQMD) Rule 1113 Architectural Coatings, rules in effect on Jan 1, 2004.）

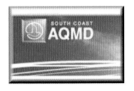

圖 9-11　南岸空氣品質管理局

- 用於室內裝修的膠合板材與植物性纖維產品，不得含有尿素甲醛樹脂（urea-formaldehyde resins）。（Composite wood and agrifiber products used on the interior of the building (i.e., inside the weather-proofing system) must contain no added urea-formaldehyde resins.）

圖 9-12　膠合板材料類

綠色清潔（Green Cleaning）

綠色清潔程序（Green Cleaning Program）

- EQ 得分項目：採購清潔產品及材料（EQ Credit: Green Cleaning－Purchase of Cleaning Products and Materials）
  - 採購綠色清潔產品：採購得到 Green Seal 綠色標章或 Environmental Choice 認證的清潔產品（Green Seal and Environmental

Choice）

- EQ 得分項：採用綠色清潔設備（EQ Credit: Green Cleaning—Cleaning Equipment）
  - 採購得到 Seal of Approval/Green Label 認證的清潔設備（Seal of Approval/Green Label）
  - 噪音規範遵循 ISO 11201 規範（Noise: ISO 11201）

9.2 最低室內空氣品質性能 / 增加通風（Minimum IAQ Performance / Increased Ventilation）

室內空氣品質（Indoor Air Quality）

維護室內空氣品質可以從以下方向著手：（Maintaining indoor air quality can be addressed through the following approaches:）

- 室內照明設計（Interior Lighting Design）
  安裝照明控制及採用照明品質設計。（Install lighting controls and adopt lighting quality design.）
- 維護建造期間的空氣品質（Protect air quality during construction）
  保護工人和設備免受灰塵和濕氣破壞。（Protect worker and equipment from dusts and moisture damage.）
- 進行大量換氣（Conduct a flush out）
- 提供大量的空氣，除去來自施工活動的汙染物。（Deliver large volume air to remove contaminants from construction activity.）
- 執行綠色清潔計畫。（Employ a green cleaning program.）

為了達到室內空氣品質的要求，LEED 對於其通風有下列的要求：

- 需符合 ASHRAE 62.1 Section 4–7, 的最低需求，通風達標的室內空氣品質（Meet the minimum requirements of ASHRAE 62.1 Section 4

－7, Ventilation for Acceptable Indoor Air Quality）

- 應遵循通風換氣率規範或適用當地法規二者中較嚴格的規定，例如加州 Title 24。（Mechanical ventilation systems must be designed using the ventilation rate procedure or the applicable local code, whichever is more stringent.(i.e. Title 24 in California State)）

9.3 禁止吸菸或車速怠速（No Smoking or Vehicle Idling）

策略（Strategies）

1. 禁止吸菸（Prohibit smoking）

吸菸對你或其他人的健康有不良的影響。（Smoking is not good for health, yours and others.）

2. 確認適當的通風（Ensure adequate ventilation）

達到空氣品質的關鍵是具備有足夠的新鮮空氣。（Sufficient fresh air is the key to achieve air quality.）

3. 監測二氧化碳（Monitor carbon dioxide）

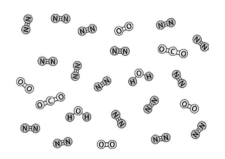

圖 9-13　空氣中之成分

- 環境菸害控制（(Environmental Tobacco Smoke) ETS Control）：禁止在建築物室內吸菸。（Prohibit smoking in the building.）

- 指定室外吸菸區，需離空調進氣口和可開窗戶必須距離入口至少 25 英尺（約 8 公尺）。（Provide signage to allow smoking in designated areas, Prohibit on-property smoking within 25 feet of entries, outdoor air intakes and operable windows.）

圖 9-14　禁菸標誌

- 在建築物出入口 10 英尺（3 公尺）貼禁菸標語。（Post signage within 10 feet（3 meters）of all building entrances.）

9.4 居住者體驗（Occuant Experience）

- 良好照明與空調。（Good lighting and HVAC.）
- 良好的照明與空調操作習慣能夠節約能源 , 也就是節約能源成本。（operation practice reduce energy use, and energy cost.）
- 增加建築物使用者對空調和照明系統控制能力，進而節約能源。（Increased users’- access to daylight can reduce the need for electrical lighting during daytime hours, thereby decreasing energy use.）
- 大部分的人會喜歡在能夠自行控制室內環境條件的地方生活、工作與玩樂。（Most of people preferred to live, work, and play in space where they have a degree of control over indoor environment.）

可以增加居住者對環境控制的方式：（Ways to increase occupant control over their environment:）

- 使用自然光線（Use daylighting）
自然採光可以節約能源和舒適居住者。
（Natural lighting can conserve energy and comfort occupants.）

- 安裝可開的窗戶（Install operable window）
打開窗戶是得到新鮮空氣最好的方式。
（Open window is the best way for fresh air.）

- 使用者調控溫度和通風（Give occupant temperature and ventilation control）

圖 9-15　自然光線

當可開啓窗口不可用時，居住者應能控制空調系統。（When operable window is not available, occupants should be provided the controls of HVAC.）

- 至少90%的居室面積裡由景觀窗部分直接看到室外景觀。（Achieve a direct line of sight to the outdoor environment via vision glazing of all regularly occupied areas.）

圖 9-16　視野良好的空間

- 提供個人舒適空調控制系統，讓 50% 以上的建築物使用者可依個人工作需求與喜好進行舒適度調整。（Provide individual comfort controls for 50%（minimum）of the building occupants to enable adjustments to meet individual needs and preferences.）

- 在多人使用的空間裡提供可調整舒適度之空調控制系統，讓使用團體可依工作需求與喜好進行舒適度調整。（Provide comfort system controls for all shared multi-occupant spaces to enable adjustments that meet group needs and preferences.）

控制系統：熱舒適（Controllability of Systems: Thermal Comfort）

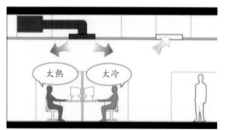

天花板送風系統

地下出風口系統

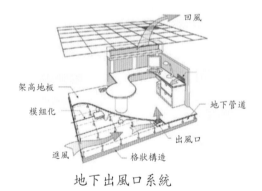

地下出風口系統

地下出風口系統

圖 9-17　室內出風系統

9.5 通用性與包容性（Accessibility and Inclusion）

策略（Strategies）

支持居住者的多樣化需求，增加建築物的廣泛可用性，以促進個人和集體的歸屬感。（To support the diverse needs of occupants and increase widespread usability of the building to foster individual and collective sense of belonging.）

符合當地通用性法規（Comply with Local Accessibility Codes）

所有專案必須通過符合 IPp：人類影響評估中確定的所有當地適用的通用性法規的設計來支持身體殘障人士的通行。如果沒有相應的法規，則需包含以下策略：（All projects must support access for those with physical disabilities through designs meeting all locally applicable accessibility codes identified in IPp: Human Impact Assessment. If there is no code in place, include the following strategies:）

1. 無障礙路線或經常使用的外部建築入口配備坡道以適應高差變化。（Accessible routes or regularly used exterior building entrances have ramps to accommodate elevation change.）

2. 所有供人通行的門都需有至少 32 英寸（0.86 公尺）的淨寬度。（All doors meant for human passage have a minimum clear width of 32" (0.86 meters).）

3. 接待處、安檢櫃台和服務櫃台都需有可供輪椅正面進入的無障礙部分。（Reception desks, security counters, and service counters all have a front approach wheelchair accessible section.）

包含至少 **10** 項與專案最相關的通用性和包容性策略：（Include at least 10 of the following accessibility and inclusion strategies most relevant to the project:）

空間多樣性通用性（Accessibility for Physical Diversity）

- 在所有經常使用的建築入口提供感應開門器或垂直手／腳按壓開門裝置。（Provide wave-to-open or vertical hand / foot press door operators at all regularly used building entrances.）

- 設計會議空間以適應至少 10% 居住者的行動輔具需求。（Design meeting spaces to accommodate mobility devices for at least 10% of occupants.）

- 健身設施中應融入無障礙且包容的設備和活動，確保有開放且無障礙的路徑通往設備周圍。（corporate accessible and inclusive equipment and activities in fitness facilities. Ensure an open and accessible route to and around the equipment.）

- 在設有非無障礙路線（如樓梯）的地方，提供一條從相同起點到終點的無障礙替代路線。（Where non-accessible routes are provided (i.e. stairs), provide an alternate accessible route that starts and terminates at the same location.）

- 在指定的無障礙洗手間或家庭洗手間內設置至少一個成人更衣站或更衣桌。（Include at least one adult changing station or table in a designated, accessible restroom or family restroom.）

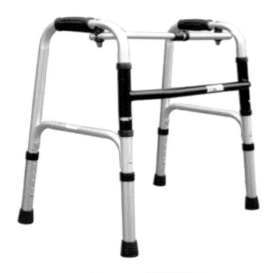

圖 9-18　　無障礙輔具　　　　圖 9-19　　無障礙路徑指示牌

安全和高齡化通用性（Accessibility for Safety and Aging）

- 提供防滑地板。（Provide non-slip flooring.0）
- 固定地毯並在所有邊緣處提供過渡條。（Fix area rugs to floor below and provide transition strips at all edges.）
- 除私人住宅外，在所有全高度玻璃處提供視覺提示或欄杆。（Provide visual indication or railing at all full height glazing, except in private residences.）
- 提供應急時的聲音和視覺警報。（Provide audible and visual alerts for emergency alerts.）

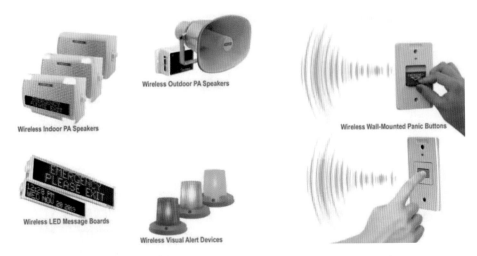

圖 9-20　緊急警報系統（Emergency Notification Systems）

- 所有樓梯需提供封閉的台階（視覺和物理上）。（Provide closed risers (visually and physically) in all stairs.）
- 在牆壁和地板、牆壁和門以及牆壁和櫃體之間使用視覺對比。（Use visual contrast between walls and floors, walls and doors, and walls and casework.）
- 在樓層高差處提供視覺、觸覺、對比色或自發光的警告。（Provide visual, tactile, contrasting, or photoluminescent warnings at floor level changes.）

圖 9-21　防滑地毯／墊（non-slip flooring）

社會健康通用性（Accessibility for Social Health）

- 提供哺乳室或哺乳艙的空間。（Provide lactation rooms or space for lactation pods.）
- 在建築的每層樓至少提供一間完全無障礙的全性別單人洗手間或一間多用途全性別洗手間。（Provide at least one fully accessible all-gender single-use restroom OR one multi-use all-gender restroom on each floor of the building.）
- 在指定的無障礙洗手間或家庭洗手間內設置至少一個成人更衣站或更衣桌。（Include at least one adult changing station or table in a designated, accessible restroom or family restroom.）
- 提供以當地人口中超過 5% 人口使用的所有語言的標誌。（Provide signage in all languages spoken by more than 5% of the local population.）

- 獲得 **EQc**：居住者體驗：生物親和環境下的至少 1 分。（Achieve at least 1 point under EQc: Occupant Experience, Option 1. Biophilic Environments.）

圖 9-22　無障礙廁所（accessible restroom）

來源：Dolphin Solutions

指示通用性（Accessibility for Navigation）

- 提供清楚標明出口、入口和主要功能的指示標誌。（Provide way-finding signage that clearly indicates exits, entrances, and major functions in the project.）
- 在標誌上提供非文字圖示和符號。（Provide non-text diagrams and symbols at signage.）
- 提供盲文、視覺和聽覺提示，及／或在行進路線上提供連續線性指示。（Provide Braille, visual and auditory cues, and/or continuous linear indicators on paths of travel.）
- 使用圖案和色塊來識別關鍵進入空間。（Use pattern and color blocking to identify key access spaces.）

- 提供觸覺／立體地圖用於指引。（Provide haptic/tactile maps for wayfinding.）

9.6 韌性空間（Resilient Spaces）

策略（Strategies）

1. 使用者照明控制（Give occupant temperature and ventilation control）

照明控制在個人工作區可以改善室內環境品質，讓居住者參與節能減排。（Lighting control in Individuals workstation improve indoor environment quality and allow occupants to participate energy conservation.）

2. 考慮必要的隔音（Consider acoustic）

噪音會降低室內工作和生活的品質，需要小心注意。（Noise reduce indoor working and living quality and needs to be taken care.）包括在 LEED-學校和 LEED-醫療保健類型，特別著重噪音之控制（Included in LEED-School and LEED-Healthcare）

3. 進行滿意度調查（Conduct occupant survey）

- 居住者的意見幫助改善室內環境品質的問題。（Occupants' opinions help improving IEQ issues.）
- 提供個人照明控制開關系統，讓 90% 以上的建築物使用者可依個人工作需求與喜好進行照明調整。（Provide individual lighting controls for building occupants to enable adjustments to suit individual task needs and preferences.）

圖 9-23　室內照明品質

- 在多人使用的空間裡（例如教室與會議室）提供可分別控制之照明系統，讓使用團體可依工作需求與喜好進行照明調整。（Provide lighting system controls for all shared multi-occupant spaces to enable adjustments that meet group needs and preferences.）

- 設計優良之空調系統與建築外殼，使室內溫熱環境能符合 ASHRAE Standard 55。（Design heating, ventilating and air conditioning (HVAC) systems and the building envelope to meet the requirements of ASHRAE Standard 55-2004, Thermal Comfort Conditions for Human Occupancy.）

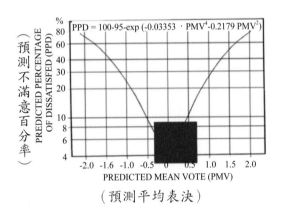

圖 9-24　PMV 滿意度調查

4. 影響熱舒適的因子（Factors affecting the thermal comfort index）

影響熱舒適指標的因素，包括室內環境因素和人為因素，如下所示。（Factors affecting the thermal comfort index, including indoor environmental factors and human factors, as follows.）

室內環境因素（Indoor Environmental Factors）：

1. 溫度（Temperatrie）

2. 濕度（Humidity）

3. 風速（Air speed）

人體因素（Human Factors）：

1. 衣著量（Clothing）

2. 活動量（Activity）

人體新陳代謝量

人體活動類型	新陳代謝活動	
	W/m²	met
平躺	46	0.8
靜坐	58	1.0
坐著從事桌面活動	70	1.2
站著輕微活動	93	1.6
站著中度活動	116	2.0

其中：1 met = 58.2 W/m²

- 在建築物人員進駐後第 6 個月至第 18 個月期間對人員進行不記名環境舒適度調查，調查使用者對於建築物環境性能之滿意程度，並找出舒適度相關問題。（Agree to conduct a thermal comfort survey of building occupants within 6 to 18 months after occupancy.）

- 倘若環境舒適度調查報告中有超過 20% 的人員對環境舒適度有所不滿，則必須承諾提出環境舒適度改善計畫。（Agree to develop a plan for corrective action if the survey results indicate that more than 20% of occupants are dissatisfied with thermal comfort in the building.）

9.7 空氣品質測試與監控（Air Quality Testing and Monitoring）

室外空氣監測（Outdoor Air Monitoring）

1. 高人員密度區域（Densely occupied area）

- 於所有高人員密度（每 1,000 平方英尺人員密度大於或等於 25 人）區域內裝設二氧化碳濃度監測設備，二氧化碳監測設備應置於離地 3 到 6 英尺的高度。（Monitor CO_2 concentrations within all densely occupied spaces (25 people or more per 1,000 square feet). CO_2 monitors must be between 3 and 6 feet above the floor.）

2. 非高人員密度區域（Non densely occupied area）

- 非高人員密度區域所使用之機械通風系統，必須裝設直接外氣流量計以測量最低外氣率。（Provide a direct outdoor airflow measurement device to measure minimum air flow in mechanical ventilation systems for non-densely occupied spaces.）

3. 安裝高效能空氣濾網（Install high-efficiency air filter）

該濾網會阻止汙染進入室內面積。（The way to stop contamination entering indoor area.）

安裝高效能空氣濾網（Install High-Efficiency Air Filter）

- 在使用機械通風的建築物中，所有居室應於人員進駐前更換空調設備濾網，最小效率值（MERV）需為 13 以上。（In mechanically ventilated buildings, install new air filtration media in regularly occupied areas prior to occupancy; these filters must provide a minimum efficiency reporting value (MERV) of 13 or higher.）

圖 9-25 濾網

- 儘可能減少汙染物質進入建築物內的機會，以防止一般使用空間發生交叉汙染。（Minimize and control the entry of pollutants into buildings and later cross-contamination of regularly occupied areas through the following strategies.）

- 在建築物所有的入口裝設至少 10 呎（300 公分）深之除塵地墊，以防止灰塵與汙染物經由人員帶入建築物內。

圖 9-26 地墊

（Employ permanent entryway systems at least 10 feet long in the primary direction of travel to capture dirt and particulates entering the building at regularly used exterior entrances.）

9.8 LEED 既有建築與 LEED 新建建築（LEED-EB vs. LEED-NC）

LEED 新版將綠色清潔這部分也納入室內環境品質的得分意涵中。（The new version of LEED has incorporated green cleaning into the indoor environmental quality scoring criteria.）

名詞解釋（Glossary）

- 氣體逸散（Off-gassing）

  合成和天然產物之揮發性有機化合物的排放。（The emission of volatile organic compounds from synthetic and natural products.）

- 病態建築症候群（Sick building syndrome (SBS)）

  主要由於室內空氣品質污染導致使用者出現各種不適症狀等問題。可能的症狀包括噁心、頭痛、咳嗽等症狀。（A combination of symptoms, experienced by occupants of a building, that appear to be linked to time spent in the building but cannot be traced to a specific cause. Complaints may be localized in a particular room or zone or be spread throughout the building. (EPA)）

- 高密度居室空間（Densely occupied space）

  每 1,000 平方英尺（每人 40 平方英尺以下）設計人員密度的區域有 25 人以上。（An area with a design occupant density of 25 people or more per 1,000 square feet (40 square feet or less per person).）

- 環境菸害 / 二手菸。（Environmental tobacco smoke (ETS), or secondhand smoke.）

空中懸浮粒子包括直接從香菸、煙斗、雪茄和間接吸菸者呼出的空氣。

（Consists of airborne particles emitted both directly from cigarettes, pipes, and cigars and indirectly, as exhaled by smokers.）

- 揮發性有機化合物（Volatile organic compounds (VOCs)）

揮發性有機化合物（Volatile Organic Compounds/VOC）是一種在常溫常壓下，具有高蒸氣壓和易蒸發性能的有機化學物質。揮發性有機物汙染多屬於逸散性排放，倘若被散發於環境中，逸散至大氣後經陽光照射，會與氮氧化物產生臭氧濃度上升及光化學煙霧等環境汙染問題，便會對人體健康造成潛在的威脅，包括刺激眼睛和呼吸管道、頭疼等，並且，當中某些化學品如苯、甲苯、鹵化碳鹵代烯烴（三氯乙烯、二氯乙烯）等已被懷疑或確定為致癌物質。揮發性有機物的來源主要為石油產品、化學溶劑、汽車尾氣和燃燒廢氣。石油產品主要存在於石油、化工、加油站等生產和銷售單位，而化學溶劑則與每個人的生活密切相關，無論是紡織品、鞋類、化妝品、油漆、傢俱、辦公用品、室內裝飾還是電子電器等設備，都可能產生揮發性的有機物質。電子產品在使用中常常會有部分的零元件會處於高溫，在此加溫狀態上容易逸散出甲醛等揮發性有機物質（VOC）的異味。（Compounds that are volatile at typical room temperatures. The specific organic compounds addressed by the referenced Green Seal Standard (GS-11) arc identified in EPA Reference Test Method 24 (Determination of Volatile Matter Content, Water Content, Density Volume Solids, and Weight Solids of Surface Coatings), Code of Federal Regulations Title 40, Part 60, Appendix A.）

題目（Practice Questions）

1. 根據環保局，美國人在室內的時間大約為多少比例？（According to the Environmental Protection Agency, what percentage of time do Americans spend indoors?）

   a. 75%

   b. 65%

   c. 90%

   d. 50%

   據 EPA 估計，美國人花費的時間 90% 在室內，其中有害汙染物的濃度可能是高危險的。（The EPA estimates that Americans spend 90% of their time indoors, where concentrations of harmful contaminants may be dangerously high.）

2. 熱舒適度通常歸因於什麼環境因素？（Thermal comfort is typically attributed to what environmental factors?）

   a. 溫度、濕度和空氣速度（Temperature, humidity, and air speed）

   b. 通風、溫度和日光（Ventilation, temperature, and daylight）

   c. 濕度、通風和可控制度（Humidity, ventilation, and controllability）

   d. 密度、溫度和太陽熱得（Density, temperature, and solar heat gain）

   ASHRAE 標準 55 的熱舒適環境因素定義為（ASHRAE Standard 55 defines the environmental factors of thermal comfort as）

   • 濕度（humidity）、空氣速度（air speed）溫度（空氣溫度和輻射溫度）（and temperature (air temperature and radiant temperature).）。

3. 縮寫 VOC 是指？（The abbreviation VOC refers to?）

   a. 揮發性有機化合物（Volatile organic compounds）

b. 變化的操作情況（Variable operating conditions）

c. 可變臭氧汙染物（Variable ozone contaminants）

d. 多功能複合組織（Versatile organized composites）

揮發性有機化合物（VOC）是在室溫下揮發。他們大多都對人體有害。
（Volatile organic compounds (VOCs) are volatile at room temperature. Many
of them are harmful to humans.）

4. 什麼策略可以改善室內空氣品質？（Which strategy supports improved indoor air quality?）

a. 消除製冷劑與臭氧消耗潛力。（Eliminate refrigerants with ozone depleting potential.）

b. 避免使用的產品具有較高的二氧化碳濃度。（Avoid the use of products with high carbon dioxide concentrations.）

c. 使用過濾器具有較低的最低效率報告值（MERV）的評級。（Use filters with a low minimum efficiency reporting value (MERV) rating.）

d. 使用回收的物料當材料（Use materials with recycled content.）

e. 使用先進的構造技術。（Use advanced framing techniques.）

f. 設計系統能夠提供充足的室外空氣。（Design systems to deliver ample outside air.）

提供大量的外部空氣進入建築物內部稀釋室內空氣汙染物，從而改善室內空氣的品質。（Delivering high volumes of outside air into the building interior dilutes indoor air contaminants, thereby improving indoor air quality.）

5. 有一個學校的前期設計方案要納入建築策略，以最大限度地提高學生的學習。下列哪個策略應被視爲實現這一目標的方法？（選擇兩個）（A school project in predesign would like to incorporate building strategies to maximize student learning. Which strategies should be considered to achieve

this goal (select two)?）

a. 安裝獨立的熱舒適（Install individual thermal comfort）

b. 引入日光進到教室（Incorporate daylight into classrooms.）

c. 減少建築能源使用低於基線標準。（Reduce the energy use of the building below the baseline standard.）

d. 考慮核心學習空間的聲學問題。（Consider acoustical issues in core learning spaces.）

b. 研究表明，在課堂上改善日光下提高學生的學習，有一項研究顯示了數學進展快 20%，閱讀進展快 26%。高性能聲學培育有效的教師與學生及學生與學生的溝通。（d. Studies show that improved daylight in classrooms increases student learning, with one study showing 20% faster progression in math and 26% faster progression in reading. High-performance acoustics foster effective teacher-student and student-student communication.）

6. 可操作的窗戶被認為是什麼類型的控制？（An operable window is considered what type of control?）

a. 照明控制（Lighting control）

b. 熱舒適控制（Thermal comfort control）

c. 聲學控制（Acoustical control）

d. 環境菸害控制（Environmental tobacco smoke control）

b. 可操作的窗戶允許居住者進行調整建築物室內之空氣的速度和溫度，由此控制多種環境條件的熱舒適。（Operable windows allow occupants to make adjustments to the air speed and temperature within the building, thereby controlling multiple environmental conditions for thermal comfort.）

7. 按需要控制的空調通風系統通常因為下列哪一個原因回應？（Demand-controlled ventilation is typically adjusted in response to __?）

   a. 時間表（A time schedule）

   b. 居住者要求建築的操作者（Occupant requests to the building operator）

   c. 二氧化碳濃度（Carbon dioxide concentrations）

   d. VOC 濃度（VOC concentrations）

   按需要控制的空調通風系統調控室外空氣進入建築物的比率。在許多情況下，建築物自動化系統檢測到室內二氧化碳濃度增加時，會增加室外氣流進入室內，以達到換氣效果。（Demand-controlled ventilation modulates the delivery of outdoor air into the building based on occupancy. In many cases, outdoor airflow is increased when the building automation system detects increased carbon dioxide concentrations.）

8. 除了禁止在建築物內吸菸，哪裡也是禁止吸菸重要的地方，以減少居住者暴露在有害化學物質的空氣中（選擇兩個）？（In addition to prohibiting smoking within the building, where is it also important to prohibit smoking to reduce occupant exposure to harmful airborne chemicals (select two)?）

   a. 在地下停車空間（In covered parking spaces）

   b. 在建築物的入口旁（Near building entrances）

   c. 鄰建築物的進氣口（Adjacent to building air intakes）

   d. 在樹林區（In wooded areas）

   b. 在靠近建築的入口或靠近建築的進氣口吸菸會使煙霧滲透至室內，讓居住者暴露在有害化學物質的空氣中。（c. Smoking near building entrances or near building air intakes allows infiltration of environmental tobacco smoke into the biding interior, thereby exposing occupants to harmful airborne chemicals.）

9. 哪種類型的產品將減少清潔用品對環境的影響？（Which type of products would reduce the environmental effects of cleaning products?）

a. CRI

b. Green-e

c. Green Label Plus

d. Environmental Choice

Environmental Choice 適用於綠色環保清潔產品中的許多標準。（Environmental Choice has many standards that apply to green cleaning products.）

10. 在哪裡吸菸是應該被禁止的？（選擇三個）（Where should smoking be prohibited for a project?(Choose 3)）

a. 可開啓的窗戶（**Operable windows**）

b. 建築入口（**Building entrances**）

c. 人行道（Walking trails）

d. 車庫（Garages）

e. 進氣口（**Air intakes**）

LEED 要求在所有新建專案禁止吸菸。吸菸不應該發生在周圍的建築物開口的地方，因爲會讓煙霧進入大樓。（LEED requires prohibiting smoking in all BD+C projects. Smoking should not occur around any building opening that could allow the smoke to enter the building.）

第十章 創新與設計（Innovation in Design）

學習目標（Learning Objectives）

- 使用創新的手法規劃設計（Using innovative approaches in planning and design）
- 將既有的表現性能更加提升（Enhancing existing performance）
- 了解 LEED 得分會根據在地化調整（Understanding localized LEED scoring adjustments）

　　本章節為 **LEED** 較為彈性之部分，可以從以下幾個方向取得分數（There are 3 primary ways to earn ID Credits）

　　IDC1：最多 5 分（ID Credit 1: (up to 5 Possible Points)）

10.1 區域優惠（Regional Priority）

　　提供符合某些特定地理環境條件之地區獎勵分數。（Regional Priority provide bonus points for project teams attempting LEED credits that address specific environmental priorities in the project's region.）

Process（過程）

檢查專案的郵政編碼是否在 RP 得分內（Check the project's zipcode for the specific RP credits）　→　調整得分目標（adjust the credit goal）　→　自動授予得分（Points are awarded automatically）

圖 10-1　USGBC 網站查詢 RP 得分

國家：美國
郵遞區號：94118
評估系統：LEED 2009 for New Construction

View Detaits:

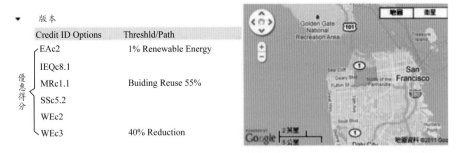

Please note: Project teams must meet or exceed the threshold liste to
earn an RPC point

https://www.usgbc.org/RPC/RegionalPriorityCredits.aspx?CMSPageID=2435

圖 10-2　查詢到特定地區顯示有優惠之項目

10.2 創新設計（Innovation in Design）

使用量化且並不在 LEED 評分系統中的既有技術或策略。（Demonstrating a Quantifiable environmental benefit using an approach not found in the LEED Rating System.）

創新表示必須比既有的 LEED 評估系統有更創新的表現。（Innovation means exceeding green building or operating goals in a significant, measurable way that is not covered in the LEED rating system.）

- 必須表現出可被量化的環境效益（Must Demonstrate a Quantifiable environmental benefit）
- 必須被廣泛的用於整個專案（Must be applied comprehensively throughout the project）
- 必須適用於不同的專案（Must be Transferable to other projects）
- ID 分數的適用性是針對個案評估的（ID Credits are evaluated on a project by project basis）
- 審查不通過時，可以採用其他手法，或直接跳過（If the ID credit is denied, the project team could submit another ID credit or just skip it）
- 設計與施工文件合併送審（Combined Review）

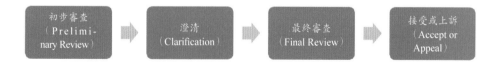

- 設計與施工文件分開送審（Split Review）

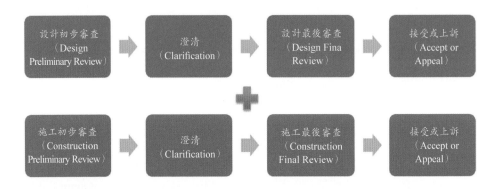

範例（Examples）

- 教育推廣計畫（Educational Outreach Program）
- 綠色清潔（Green Housekeeping）
- 提高混凝土爐灰比（High Volume Fly Ash）
- C2C 綠色環保建材（Purchase of Credle-to-Credle products）
- 廚餘處理系統（Food Compositing facility）
- 達到其他評估系統的得分項目需求（Achieve LEED Credit from a different rating system）

10.3 優異表現（Exemplary Performance）

超越在 LEED 認證系統中列出的得分要求。（Exceeding the Credit requirements listed in LEED Rating System.）

- 試驗性的分數（最多 1 分）（LEED Pilot Credit (Max. 1 point)）
  嘗試 LEED 系統中實驗性質的得分。（Try the tested credits or prerequisites that may be used in upcoming versions of the LEED rating systems.）

優異表現（ID Credit 1 Exemplary Performance）

　　若建築物的設計或性能超越 LEED 評估系統內容所訂定之需求，則可由此獲得額外的分數。（Project teams can earn exemplary performance points by implementing strategies that result in performance that greatly exceeds the level or scope required by an existing LEED prerequisite or credit.）

- 給予下一個增量更高級別的分數（Awarded for going to the next higher incremental level of credit performance）

  如果分數百分比為 10% 和 20%，而下一個層級有可能會是 30%（(ie. If the credit threshold percentage required 10% and 20%, 30% would be the next threshold)）

- 給予加倍等級的信用性能（即如果目前得分值百分比是 50%，可能 100% 可以賺取 EP 的分數）。（Awarded for doubling the level of credit performance. (ie. If the credit threshold percentage required 50%, then 100% can earn an ID point).）

- 它不適用於每個 LEED 得分項目（It is not available for every LEED Credit）

範例（Examples）

- 日常節水 55% 以上，可多得 1 分（if a project demonstrates a 55% reduction in potable water use）

用水量減少百分比 （Water Use Percent Reduction）	分數 （Points）
25%	1
30%	2
45%	3
40%	4

用水量減少百分比 （Water Use Percent Reduction）	分數 （Points）
45%	5
50%	6

其它可以獲取 EP 得分的項目：

- 增加使用替代交通運輸（Increasing the use of alternative transportation）

- 雨水管理（Rainwater management）

- 減少熱島效應（Reducing the heat island effect）

- 減少室內用水（Reducing indoor water use）

- 增加能源效能（Increasing energy performance）

- 提高現場可再生能源（Increasing on-site renewable energy）

- 增加綠色能源使用（Increasing green power use）

- 增加使用環保材料（Increasing the use of environmentally preferable materials）

LEED 試行得分圖書館（LEED Pilot Credit Library）

試行得分正在測試 LEED 的新版本分數。團隊可以嘗試在創新類這些得分。美國綠色建築委員會使用團隊的反饋，以確認接下來的 LEED 版本所使用是否合適。而這些項目可以在 USGBC 的網站查詢。（Pilot credits are credits that are being tested for an updated version of LEED. Project teams can attempt these credits under the Innovation category. The USGBC uses project teams' feedback on these credits to determine if the credits will actually be used in upcoming version of LEED.）

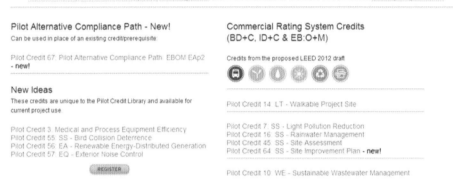

圖 10-3　LEED Pilot 試行得分畫面

　　專案團隊中至少要指派有一位取得 LEED 認證之專業人員（LEED AP）協助專案執行，LEED AP 不但可以協助 LEED 綠建築專案所要求的高度設計整合性，並使申請與認證過程順利進行。（LEED Accredited Professionals have the knowledge required to integrate sustainability into building operating and upgrade practices. The LEED Accredited Professional understands the importance of considering interactions between the LEED prerequisites and credits and their respective criteria.）

必須的提交資料：（Required submittal information:）

- 在 LEED AP 的名稱（Name of the LEED AP）
- LEED AP 的公司名稱（Name of the LEED AP's company）
- LEED AP 證書影本（Copy of the LEED AP certificate）

10.4 LEED 認證之專業人員參與（LEED Accredited Professionals）

在專案團隊裡有 LEED AP。（Having a LEED AP on the project team.）

在專案團隊成員中有 LEED AP 一位即可取得 1 分，但若有兩位 AP 也只能取得 1 分。

題目（Practice Questions）

1. 一個專案團隊正在考慮追求創新設計得分的策略，這將需要大量的成本。該小組決定，要執行一措施以獲得創新設計的得分，團隊應該採取什麼措施，以確定其策略是否將有資格獲得分數？（A project team is considering pursuing an Innovation in Design credit for a strategy that will require a large capital expense. The team decides that it will implement the strategy only if the strategy will be eligible for an Innovation in Design credit. What steps should the team take to determine whether its strategy will be eligible for the credit?）

   a. 研究現有的用戶，看有先前已發表之措施。（**Research existing CIRs to see whether the strategy has been previously addressed.**）

   b. 計算整體環境影響和自我評估的策略是否有顯著。（Calculate the strategy's overall environmental impact and self-evaluate whether it is significant.）

   c. 草擬有說服力的說明提交給 LEED。（Draft a compelling narrative for the LEED submittal.）

   d. 檢查 LEED 的評分系統。（Check the LEED rating system.）

a. 符合條件的創新策略是不是在 LEED 評級系統或參考指南之內處理，但透過 CIR 過程代替評估。如果某一問題從未被定義過，通常透過 CIR 詢問。（Eligible innovative strategies aren't addressed within the LEED rating system or reference guides, but are instead evaluated via the CIR process. If the issue in question has never been addressed, a CIR is typically required to evaluate.）

2. 專案如何識別哪些積分有區域優先得分？（Projects can identify which credits have regional priority by?）

   a. 在美國綠色建築協會網站上查詢（Checking the USGBC website）

   b. 研究當地相關的環境議題（Researching locally pertinent environmental issues）

   c. 提交信用解釋要求（CIR）（Submitting a Credit Interpretation Request (CIR)）

   d. 計算淨環境影響（Calculating the net environmental impact）

   a. 美國綠色建築協會可透過郵遞區號查詢，來了解各地區域之優先得分，可從該組織的免費網站上下載列表。（USGBC maintains a listing of all regional priority credits by ZIP code, which can be downloaded from the organization's website free of charge.）

3. 在創新策略的創新設計得分很可能透過專案的什麼作為取得？（An Innovation in Design credit for innovative strategies is likely available for projects that?）

   a. 發出新聞稿，宣布自己的 LEED 專案已登記（Issue a press release to announce their LEED project registration）

   b. 包含二個 LEED 認可專業人士在專案團隊之中（Include two LEED Accredited Professionals on the project team）

c. 建立一個教育推廣方案（**Develop an educational outreach program**）

d. 計畫註冊在 LEED 既有建築專案：營運和維護（Plan to enroll the project in LEED for Existing Buildings: Operations and Maintenance）

c. 創新策略，擴大綠色建築實踐的廣度和引入新的思路。他們和卓越表現是不同的，因為它們解決的環境問題不是 LEED 評級系統有包含的。教育推廣計畫的目的是教育有關建築物的功能和綠色建築的總體效益。這些概念都沒有在其他的 LEED 評級系統內被提到，因此這種策略因為它設計的創新而有可能成為創新策略。（Innovative strategies expand the breadth of green building practices and introduce new ideas. They are distinct from exemplary performance innovation credits in that they address environmental issues not addressed elsewhere within the LEED rating system. Educational outreach programs are intended to educate building occupants about the building's features and the benefits of green building in general. These concepts aren't addressed elsewhere within the LEED rating system, so this strategy is a likely candidate for an Innovation in Design credit for innovative strategies.）

第十一章　LEED GA 認證考試模擬試題及詳解（LEED GA Practice Questions）

11.1 模擬考試題（LEED GA Practice Questions）

1. 在 LEED® 專案中甚麼樣的情況可以允許使用氟氯化碳（CFC）冷媒？

   A. 隨時隨地

   B. 在現在既有之建築中的空調設備

   C. 如果氟氯化碳（CFC）逐步淘汰計畫是不符合經濟成本

   D. 如果製冷劑每年洩漏率低於 10%

2. 提交 CIR 的時候應該提交包含哪一部分的資訊？

   A. 照片或圖說

   B. 專案基本資訊

   C. 適用保密條款

   D. 得分項目名稱

   E. 您的連絡人資訊

3. 下列關於 CIR 流程的敘述何者正確？

   A. CIR 申請成功可以被授予 LEED 分數

   B. CIRs 每個申請的費用是 US$220

   C. 可以由美國綠色建築委員會的任何帳戶提交 CIRs 問題

   D. CIRs 必須都適用於所有專案

4. 誰可以為 LEED® for Homes 評估並執行現場視察和為住宅評分系統測

試？

A. Home 檢查員（Home Inspector）

B. 專案經理（Project manager）

C. Home 評分員（Home Rater）

D. 綠色評分員（Green Rater）

E. 驗證調試人員（Commissioning Agent）

5. 以下的法規項目哪個是和分區號碼有關？（選擇三項）

A. 景觀法規

B. 能源使用法規

C. 停車空間要求

D. 開放空間要求

E. 雨水管理法規

6. 在 LEED 評級系統中哪一個評估系統會因為已經被另一個評級系統中的 LEED 認證而得分？（選擇三項）

A. LEED 學校

B. LEED 住宅

C. LEED 室內裝修版

D. LEED 營運和維護版

E. LEED 核心及外殼版

7. 三重底線的三個組成部分是什麼？（選擇三項）

A. 社會責任

B. 環境管理

C. 持續教育

D. 政府法規

E 經濟繁榮

8.　以下哪一項不是回收的收集和儲存區所需的設計考慮必要條件？

　　A. 方便位置

　　B. 清運卡車可以到達

　　C. 適當的標示牌

　　D. 保護設施元件

9.　當總承包商收集有關建築廢棄物管理得分項目的文件，哪兩個量化的材料資料是最有用的說明文件？（選擇 2 項）

　　A. 將門窗捐助給其他回收單位的數量

　　B. 混凝土用作填充材料的重量

　　C. 廢金屬之回收的體積量

　　D. 場址清理的廢棄物製成堆肥的體積

　　E. 基地木材廢料的燃燒量

10.　熱島效應在城市地區會有比較高的溫度，而與農村相比，可能提高多少溫度？

　　A. 2 度

　　B. 4 度

　　C. 6 度

　　D. 8 度

　　E. 10 度

11.　以下哪一項不被認為是「基本服務設施」？

　　A. 銀行

　　B. 食品雜貨店或商店

　　C. 停車場

　　D. 美髮沙龍

E. 教會

12. 停車場使用開放網格鋪面的系統可以在兩個相關的永續性概念／得分項目有貢獻？（選擇 2 項）

A. 雨水管理的流量

B. 基地選擇

C. 用水量減少

D. 替代性交通、停車容量

E. 非屋頂的熱島效應

13. 適合飲用的水的兩個主要來源是什麼？（選擇兩項）

A. 收集的雨水

B. 湖泊或河流

C. 自來水系統

D. 海洋

E. 專用或公共的井水

14. 如果承包商希望對源頭控制用作建築廢料管理的一個組成部分計畫，下列哪一個策略是這樣的實踐？（選擇兩項）

A. 將材料回收整理後捐贈給其他人

B. 訂購預先組裝的產品

C. 採用模組化的結構技術

D. 回收產品包裝材料

E. 基地中再利用廢舊材料

15. 以下產品將被視為包含消費後回收的內容，將有助於 LEED 中回收材料的得分？（選擇三項）

A. 鋼箱梁由使用過的鐵軌重製

B. 玻璃地磚由在窗戶製造過程破掉的玻璃製成

C. 由回收的汽水水瓶製成的地毯

D. 由回收塑膠製成的水管套件

E. 由報紙製成的吧檯

16. 以下專案中的項目哪個被認為是「軟成本」？

A. 專案中的結構用鋼

B. 景觀設施成本

C. 建築師設計費

D. 給泥作工匠的工資

E. 地毯

17. 區域優先得分專案是基於什麼給予？

A. 專案的城市

B. 專案的狀態

C. 專案的 zip 代碼

D. 專案的縣市

18. 哪兩個特定類型的冷媒在蒙特利爾議定書中提到必須逐步淘汰使用的？（選擇兩項）

A. 天然冷媒

B. 氟氯碳化物（Chlorofluorocarbons, CFCs）

C. 氫氟氯碳（Hydrochlorofluorocarbons, HCFCs）

D. 氫氟碳化合物（Hydrofluorocarbons, HFCs）

E. 海龍（Halons）

19. 使用本地區域材料的兩個主要好處是什麼？（選擇兩項）

A. 支援當地經濟

B. 降低交通運輸的影響

C. 保證產品品質與工藝

D. 減少成本

20. 專案使用 LEED BD＋C、LEED ID＋C 和 LEED O＋M 的時候，滿足
專案的最低專案要求（MPRs）中的面積要求是多少？

A. 500 平方米／英尺

B. 1,000 平方米／英尺

C. 1,500 平方米／英尺

D. 2,000 平方米／英尺

E. 2,500 平方米／英尺

21. 以下哪一項不是 LEED 減少單獨開車與車輛減少使用的策略？

A. 提供的法定所需的最低停車量

B. 停車收費

C. 基地位址選擇於附近有公共交通之處

D. 提供自行車存放處和淋浴

E. 提供建議的低油耗車和低排放量車輛專用車位

22. 以下哪些選項被認為是自然冷媒？（選擇三項）

A. 丙烷（Propane）

B. 氟利昂（Freon）

C. 氨（Ammonia）

D. 二氧化碳（Carbon Dioxide）

E. 高純度鐵（Puron）

F. 一氧化碳（Carbon Monoxide）

23. 每年固體廢棄物中有多少的百分比是歸因於建設和拆除所浪費的？

A. 10%

B. 20%

C. 30%

D. 40%

E. 50%

24. 在 LEED 專案中，整合性設計流程所產生最大的效益是？

　　A. 縮短專案時程

　　B. 協同作業於各項得分項目

　　C. 減少專案軟成本

　　D. 了解建築基本資訊

　　E. 教育業主

25. 以下哪個選項沒有辦法減少雨水徑流的量？

　　A. 最小化建築投影面積

　　B. 增加不透水鋪面

　　C. 使用生態池

　　D. 安裝屋頂花園

　　E. 雨水採集系統

26. 以下哪個人沒有進入 LEED 線上（LEED online）的權限？

　　A. 建築師

　　B. 業主

　　C. 建築使用者

　　D. 一般承包商

　　E. 景觀建築師

27. 如果一個專案有 1 全職（每週 40 小時）員工和 6 位兼職（每週 20 小

時）員工，則此專案的全職等效（FTE）值是多少？

A. 3

B. 4

C. 5

D. 6

E. 7

28. 在 LEED 之中，下列哪個類型的車輛不能算是替代燃料車輛？

A. 氫動力

B. 電動動力

C. 油電複合（電＋油）

D. 柴油動力

E. 乙醇動力

29. 以下哪個策略並不屬於提供「優先停車位」的方式？

A. 提供停車票卡打折優惠

B. 提供優先選擇停車位的機會

C. 提供靠近建築物的停車位

D. 提供遮陽有頂的車位

E. 提供一個隨機不指定的停車位

30. 哪種類型的水不允許用於灌溉目的？

A. 飲用水（Potable water）

B. 井水（Well water）

C. 黑水（Blackwater）

D. 中水（Graywater）

E. 收集的雨水（Rainwater）

31. 如果一個專案團隊在 LEED 專案中使用中水回收系統，哪兩個是最合
　　適用來使用的用途？（選擇兩項）

　　A. 洗手

　　B. 沖洗馬桶

　　C. 洗衣服

　　D. 滴灌灌溉

　　E. 噴霧灌溉

32. 在應回收的計畫中，其中以下材料中哪些是滿足存儲和回收的前提條
　　件，必須納入的最基本項目？（選擇三項）

　　A. 玻璃

　　B. 墨水匣

　　C. 燈泡

　　D. 塑膠

　　E. 紙板

33. CRI 綠色標籤認證（Green Label Plus Certification）提供了什麼類型的
　　建築材料的證明？

　　A. 油漆

　　B. 木材產品

　　C. 地毯

　　D. 傢俱

　　E. 黏著劑

34. 在許多 LEED 評級系統的標準規定中，有關於建築物能源使用和性能
　　的法規是？

　　A. Green Seal Standard 36

B. ASHRAE 標準 52.2-1999 年

C. ASHRAE 標準 55-2007 年

D. ASHRAE 標準 90.1-2012 年

E. BIFMA International

35. 哪一個項目是 LEED 認為必須要提供完整的供應鏈產銷文件用來證明並取得分數？

A. 認證木材

B. 回收材料

C. 區域材料

D. 快速再生材料

E. 材料重複使用

36. 有哪個評分體系是某些專案可能取得 LEED 核心及殼（Core and Shell）後還可以再取得的認證？（選擇 2 項）

A. LEED 商業內裝（LEED for Commercial Interiors）

B. LEED 新建建築（LEED for New Construction）

C. LEED 學校（LEED for Schools）

D. LEED 營運和維護（LEED for Operations and Maintenance）

E. LEED 住宅（LEED for Homes）

37. 哪種類型的製冷劑基本上已是零臭氧破壞潛力（ODP），但仍具有高全球暖化潛力（GWP）？

A. 天然冷媒

B. 氟氯碳化物冷媒（CFCs）

C. 氫氟碳化合物冷媒（HFCs）

D. 含氯氟烴冷媒（HCFCs）

38. 下列哪一個評分系統必須要定期重新認證？

　　A. LEED 營運和維護版（LEED for Operations and Maintenance）

　　B. LEED 新建建築版（LEED for Building Design and Construction）

　　C. LEED 住宅版（LEED for Homes）

　　D. LEED 的室內設計和裝修版（LEED for Interior Design and Construction）

　　E. LEED 社區發展版（LEED for Neighborhood Development）

39. 在 LEED 的可再生能源評分系統下，下列哪一項不是可以接受的？（選擇 2）

　　A. 太陽能板（Photovoltaic Panels）

　　B. 地熱交換系統（Geo-exchange System）

　　C. 風力渦輪系統（Wind Turbine）

　　D. 低影響水力系統（Low-impact Hydroelectricity）

　　E. 核能發電（Nuclear Energy）

40. 以下哪一個選項的人在建築物中被視為「短期居住者」？（選擇 2 項）

　　A. 大學學生

　　B. 小學學生

　　C. 辦公室員工

　　D. 工廠工人

　　E. 商店客戶

41. 在計算生命週期成本時，通常專案團隊不會考慮下列哪一個項目？

　　A. 營運成本

　　B. 維護成本

　　C. 初始成本

D. 施工成本

E. 以上皆非

42. 哪一個評估系統要求專案團隊需要建立初步評分？

A. LEED 學校版（LEED for Schools）

B. LEED 殼心及外殼版（LEED for Core and Shell）

C. LEED 社區發展版（LEED for Neighborhood Development）

D. LEED 營運與維護版（LEED for Operations and Maintenance）

E. LEED 住宅版（LEED for Homes）

43. 如果美國綠色建築委員會發現專案不符合最低程式要求（MPRs），則專案會發生什麼？

A. 認證會被撤銷

B. 必須支付罰款

C. 專案的認證等級降低一級

D. 必須重新提交專案

44. 哪種類型的植物可以減少灌溉水需求？（選擇 2 項）

A. 侵入性植物（外來種）

B. 本地物種

C. 適應性物種

D. 栽培性植栽

45. 下列何者是 LEED 可以適當使用標誌的用途？

A. 產品外包裝

B. 連結到網站（www.usgbc.org/leed）

C. 表明組織是美國綠色建築委員會會員

D. 爲了追求 LEED 專案認證

46. 下列何者不是中水的來源？

    A. 淋浴

    B. 廁所馬桶

    C. 機器洗滌水

    D. 浴室臉盆

    E. 飲水機

47. 下列材料不可能計算於建築廢物管理原則？（選擇 2 項）

    A. 土地清理產生的廢土

    B. 廢棄木材

    C. 石棉地板、瓷磚

    D. 混凝土

    E. 產品包裝

48. 下列哪一個冷媒在一般常見冷媒中具最高的全球暖化潛力（GWP）？

    A. 氟氯化碳 -502 (CFC-502)

    B. 氟氯烴 -123 (HCFC-123)

    C. HFC-404A (HFC-404A)

    D. 氟氯化碳 -11 (CFC-11)

    E. HFC 23 (HFC-23)

49. 對一個大型辦公室建築的停車場來說，以下哪個策略不會適當的減輕熱島效應？

    A. 使用成熟的樹木遮蔭 50% 的停車場

    B. 使用太陽反射係數（SRI）35 的鋪面

    C. 至少一半的停車場空間地下化

    D. 使用針葉樹樹苗的樹蔭來遮蔭停車場

E. 使用開放網格鋪面

50. 哪一個專案部分包含「發展足跡（development footprint）」的專案範圍呢？（選擇三項）

A. 大樓本身

B. 天然池塘

C. 樹木繁茂地區

D. 景觀

E. 停車場

F. 人行道

51. 以下哪一項不是為了儘量減少非屋頂熱島效應的策略？

A. 使用開放網格鋪面

B. 用太陽能板遮蔭硬鋪面

C. 停車場地下化

D. 停車場使用瀝青材質

E. 使用太陽反射係數 29 以上的材料

52. 專案團隊使用隔熱棉花材料從種植和製造都在基地範圍內，100 英里的範圍。以下哪個項目可以得分？

A. 回收內容

B. 快速再生材料

C. 本地區域材料

D. 本地區域材料和回收內容

E. 快速可再生材料和本地區域材料

53. 「綠色標籤（Green Tags）」通常是指：

A. 任何形式的可再生能源

B. 可持續產品認證

C. 可再生能源證書（RECs）

D. 獎勵、採購省油車輛

54. 如果專案團隊工作的 LEED BD＋C 專案可以獲得材料或產品的製造文件，但是沒有開採的文件紀錄，專案團隊應該如何計算其在本地區域材料的得分？

A. 他們可能只計算 50% 的材料

B. 他們可以計算全部價值的材料

C. 他們不能計算在材料中得分

D. 只有材料被製造在距離基地 100 英里以內的範圍，專案團隊才能計算材料

55. 哪一個評分系統提供創新和設計（ID）的得分，如果有 LEED 專家委員會？

A. LEED 建築設計和施工

B. LEED 的室內設計和裝修

C. LEED 營運和維護

D. LEED 住宅

E. LEED 社區發展

56. 太陽能反射率與下列哪一個名詞有相同的意義？

A. 含水層（Aquifer）

B. 遮罩（Shielding）

C. 眩光（Glare）

D. 流明（Lumen）

E. 反射率（Albedo）

57. 減少屋頂的熱島效應有哪兩個主要策略？（選擇兩項）

    A. 不在屋頂上裝設機械設備

    B. 使用屋頂材料具有高的太陽能反射指數

    C. 使用金屬板代替礫石材料

    D. 避免使用斜屋頂

    E. 使用綠屋頂或屋頂花園

58. 以下的選項哪個不是負責的 LEED 專案管理者的權限？

    A. 邀請專案小組成員至 LEED online

    B. 管理 LEED 線上系統

    C. 給予專案的得分

    D. 提交 CIRs

    E. 註冊專案

59. 專案要如何購買「綠色能源」？（選擇兩項）

    A. 透過本地水電公司提供的方案

    B. 透過購買綠色能源認證（Green-e Certified (RECs)）

    C. 透過購買和安裝太陽能板

    D. 透過直接從附近的風力發電廠購買能源

60. 什麼類型的植物最適合使用在綠屋頂作為屋頂植栽？

    A. 一年生草本

    B. 栽培性植栽

    C. 本土性植物

    D. 樹和灌木

61. 美國鈑金與空調包商協會（SMACNA）是針對哪個部分提供規範準則？

A. 空氣品質

B. VOC 成分

C. 冷媒使用

D. 建築廢棄物管理

E. 雨水回收管理

62. 地方法規及規章可能會影響到以下哪兩個策略？

A. 建築廢棄物管理

B. 利用雨水沖洗馬桶

C. 最小化停車空間

D. 施工期間的室內空氣品質管理

E. 採購綠色電力

63. 以下哪一項是 LEED 評分系統所認為的「適當的建築基地」範圍？（選擇 2 項）

A. 優良農田

B. 具有瀕臨絕種物種的棲息地

C. 距離自然水體 50 英尺的範圍

D. 褐地

E. 已經開發過的都市基地

64. 以下哪一項不是創新和設計策略必須符合的？

A. 它必須是全面性的

B. 它必須符合經濟效益

C. 它必須可以重複使用在其他專案

D. 它必須是可量化的

E. 它必須是目前不存在 LEED 評分系統之中的項目

65. 如果 LEED 專案提供優先低油耗和低排放車輛停車的車位，停車空間必須設立及符合下列哪一項敘述？

A. 只需要提供一個優先停車空間

B. 優先停車空間只需要位於訪客來訪的停車場

C. 優先停車空間只需要設立在工作人員使用的停車場

D. 優先停車位必須分佈在所有的停車空間

66. 氟氯碳化物（CFC）和氫氟氯碳（HCFC）使用造成了什麼環境問題？

A. 水汙染

B. 空氣汙染

C. 土壤侵蝕

D. 臭氧層破壞

E. 熱島效應

67. 開放空間有甚麼樣的條件？

A. 植栽茂盛

B. 可以通往公共區域

C. 在晚上開燈

D. 是直接毗鄰建築物的入口

E. 它為瀕危物種提供棲息地

68. 哪一型的冷媒一般來說具有最好的效率？

A. 氟氯碳化物（CFCs）

B. 氫氟碳化物（HFCs）

C. 氫氟氯碳（HCFCs）

D. 天然冷媒

69. 建築廢料管理一部分計畫運用混合回收的方式，有下列哪一個潛在缺點？

A. 它比分類回收占用更多的空間

B. 它對專案團隊來說可能需要更多的時間

C. 它可能比分類回收成本更高

D. 它比分類回收更困難記錄與管理

70. 什麼措施可以防止行人進入到社區公共資源？（選擇三項）

A. 距離

B. 公共公園

C. 高速公路交會處

D. 牆

E. 停車位

71. 哪一項聯邦法規已通過並被許多當地地方政府使用，同時要求無障礙設施與空間？

A. 美國冷凍空調協會（ASHRAE）

B. 能源政策法（EPAct）

C. 公民法（Civil Rights Act）

D. 美國殘疾人法案（ADA）

E. 美國國家標準協會（ANSI）

72. 自來水的最佳用途是該如何運用？

A. 冷卻塔的冷卻水

B. 灌溉用途

C. 游泳池

D. 飲用

E. 廁所沖洗

73. 哪種策略不會有效地減少自來水於灌溉的使用？

A. 噴霧灌溉

B. 滴灌灌溉

C. 地膜覆蓋

D. 使用本機或改良性植物

E. 雨水收集

74. 下列哪一個策略是 EPA 認為減少固體廢棄物的最佳辦法？

A. 材料回收

B. 源頭減量

C. 材料重複使用

D. 材料焚燒

E. 使用回收成分的材料

75. 理想情況下，整合式設計應該於哪一個階段開始？

A. 在專案前期策劃階段

B. 當專案工程建設施工階段開始

C. 當專案已經在 GBCI 登記為註冊專案

D. 當專案通過該專案的設計階段

76. 下列哪一個專案將絕對不能滿足 LEED 的最低專案要求（MPRs）？
（選擇兩項）

A. 建立在遊船上的賭場

B. 改造和增修一間大辦公室

C. 新的小學分校

D. 臨時體育設施

E. 5000 平方英尺消防隊大樓

77. 下列哪一個設備或系統不會產生作業用水？

A. 家用熱水器

B. 洗碗機

C. 洗衣機

D. 冷卻水塔

E. 蒸鍋

78. 以下的衛生設備的用水量哪些是加侖每分鐘（GPM）作為單位？（選擇三項）

A. 馬桶

B. 廚房水龍頭

C. 淋浴蓮蓬頭

D. 浴室水龍頭

E. 小便斗

79. 下列哪一項 LEED 類別有最多的 LEED 得分比重？

A. 永續基地

B. 用水效率

C. 能源和大氣

D. 材料和資源

E. 室內環境品質

80. 爲什麼專案團隊有時候會將某個項目超過 LEED 得分項目之既定的基準值？

A. 獲得卓越性能（EP）

B. 保護環境

C. 防止得分項目被審查到

D. 爲了展現團隊的專業

81. 在 LEED 評分系統中，能源政策法（Energy Policy Act）用在何處？

    A. 設置產品 VOC 限制

    B. 建立建築能源性能的基準

    C. 以建立建築幫浦設備性能的基準

    D. 設置室內空氣品質要求

82. 哪一個是常用於分析可持續設計的成本策略方法？

    A. 計算投資回收期

    B. 內部報酬率

    C. 生命週期成本

    D. 優勢分析

83. 以下哪一項是最佳的基地開發地點？

    A. 未開發的素地

    B. 優良農田

    C. 百年洪水線範圍

    D. 以前開發過的基地

    E. 濕地

84. 哪兩個標準是用於建立與查詢低排放量和燃油效率之車輛的基準？
    （選擇兩項）

    A. 美國加州空氣資源局（ZEV）

    B. 美國經濟節能委員會的（American Council for an Energy-Efficient Economy Green Book）

    C. Greenguard 認證專案

    D. Green Seal 標準

    E. 綠色電子產品認證

85. 如果專案團隊中有十位 LEED AP，則在 LEED 專案中將取得多少分
數？

    A. 1

    B. 5

    C. 10

    D. 20

86. 透水鋪面中「開放網格」的面積必須要至少滿足多少比例？

    A. 20%

    B. 30%

    C. 40%

    D. 50%

    E. 60%

87. 在每一個 LEED 建築設計和施工（BD＋C）評分系統的必要條件中，
哪個文件是必要提出用以說明的內容？

    A. 專案說明

    B. 專案圖面

    C. 照片

    D. LEED 得分表格

    E. 所需要之簽名

88. LEED 認證收費根據：

    A. 專案面積

    B. 專案複雜性

    C. 專案成本

    D. 審查提交的數量

89. 下面哪些材料不被認爲是快速再生資源？

    A. 竹

    B. 軟木塞

    C. 稻草

    D. 絲綢

    E. 木材

90. 在 LEED 專案中材料或產品開採或加工製造和運輸距離基地必須在多少範圍以內才可以算是本地區域性材料？

    A. 100 英里

    B. 150 英里

    C. 250 英里

    D. 500 英里

    E. 1000 英里

91. 以下哪一項是任何 LEED 專案都強制要做的條件？（選擇兩項）

    A. 得分項目

    B. 最低專案要求

    C. 專案有 LEED 認證專業人員

    D. 必要條件

    E. 整合性的專案團隊

92. 以下哪一個運輸方法不能算作 LEED 所認定的公共交通？

    A. 公共汽車

    B. 地鐵系統

    C. 高架軌道系統

    D. 計程車

E. 校園班車或校車系統

93. 一個新的高層及集合住宅應該用哪一個 LEED 評分系統來認證？

A. LEED 住宅版

B. LEED 核心及外殼版

C. LEED 新建建築版

D. LEED 營運和維護

E. LEED 室內設計和裝修

94. 下列哪一個是施工活動汙染防治計畫所需要防治的項目？（選擇 3 項）

A. 侵蝕

B. 固體廢棄物

C. 建築工人事故

D. 碳排放量

E. 沉澱

F. 空氣中粉塵汙染

95. 從馬桶產生的廢水被認為是哪種類型的水？

A. 中水

B. 黑水

C. 飲用水

D. 雨水

96. 與 EPAct 所訂定出的小便斗用水的基準值相較，使用一個無水尿斗於廁所會減少多少的用水量？

A. 1.6 GPF

B. 1.0 GPF

C. 0.5 GPF

D. 2.5 GPM

97. 在全部 110 分當中，有多少的分數是屬於「加分」的分數？

A. 5

B. 10

C. 15

D. 20

98. 下列哪一個不是熱舒適性的環境因數？

A. 空氣溫度

B. 輻射溫度

C. 空氣速度

D. 濕度

E. 空間

99. 以下哪一項項目不是 LEED 營運與維護版當中所訂定的範圍？

A. 用水效率

B. 材料和資源

C. 室內空氣品質

D. 創新設計

E. 區域優先

100. Xeriscaping 對於下列哪一個部分可以貢獻得分？

A. 減少熱島效果

B. 用水高效灌溉

C. 建築廢棄物管理

D. 室內空氣品質

E. 認證木材

11.2 模擬考試題詳解

1. C. 如果逐步淘汰不可行

   LEED 專案允許氟氯化碳逐步淘汰，結果發現要投資回收期十餘年以上（對於轉換和替代方案）的專案保持他們現有的氟氯化碳系統，但是洩漏率必須低於 5% / 每年。

2. B. 專案背景資訊

   CIRs 不能包含任何附件，包括圖紙和專案資訊，他們不應包含任何有關專案的機密資訊。

3. B. CIRs 費用每個 $220

4. D. 綠色評分員

5. A. 環境法規

   C. 停車空間要求

   D. 開放空間要求

   能源使用規範通常包含在建築法規和雨水排放法規，而且都由當地汙水區管理，不是分區管理單位。

6. B. LEED 住宅

   C. LEED 商業內裝版

   D. LEED 營運維護版

7. A. 社會責任

   B. 環境管理

   E. 經濟繁榮

   三重底線通常也被稱為「人類、地球與繁榮」。

8.　B.載運卡車通路

它必須要考慮如何收集可回收垃圾的卡車，但是它不是前提所需的條件。

9.　B.混凝土用作填充材料的重量

C.廢金屬之回收的體積量

建築廢物管理得分基於重量或體積來計算，因此其他選項中的單位都是經過轉化後，並非一開始的重量或體積。

10.　E.10 度

11.　C.停車很多

12.　A.雨水管理的流量

E.非屋頂的熱島效應

開放網格鋪面因為通透面積增加，從而促進雨水滲入和開放，因此可以減輕熱島效應。

13.　C.自來水系統

E.公共或私人的井

其他選擇都不被視為飲用，因為未經處理。

14.　B.訂購預裝配的產品

C.使用模組化施工技術

而其他選項都是減少施工手段浪費，只有預裝配和模組化施工減少一開始的來源。

15.　A.鋼樑利用鐵軌再製

C.由回收的蘇打瓶製成地毯

E.由報紙製成的檯面

所有的選項包含回收的成分，但在玻璃瓷磚回收的內容將被視為消費

前的材料和塑膠回收的內容不利於 LEED 回收成分的得分。

16. C.建築師設計費

所有其他選項都被認爲硬成本。

17. C.專案的郵遞區號

18. B.氟氯碳化物（CFCs）

C.氫氟氯碳（HCFCs）

19. A.支持當地經濟

B.儘量減少運輸的影響

雖然當地材料可以是更低的成本和可能的品質非常高，這些屬性並不保證他們在本地進行。

20. B.1,000 平方英尺

21. B.收費停車

雖然收費停車可能減少一人開車的情況，它不是被LEED接受的策略。

22. A.丙烷

C.氨

D.二氧化碳

23. D.40%

24. B.協同作業於各項得分項目

而其他選項都是可能的整合化設計過程的好處，但協同作業於各項得分項目是最大的 LEED 具體好處。

25. B.增加不透水鋪面

增加透水面積將實際上增加地表的雨水徑流，而不是減少。

26. C. 建築使用者

   大廈的業主、設計師和建設者需要獲得 LEED 線上檔的專案，建設使用者不需要訪問。

27. B. 4

   (1)全職員工 = 1 FTE，6 兼職員工（20 小時／週）= 3 FTE、1 FTE + 3 FTE = 4 FTE

28. D. 柴油動力

29. E. 提供隨機分配的車位空間

   提供分配的空間最接近建築物可稱為優先的停車位，但不是只是隨機分配的空間。

30. C. 黑水

   黑水公司被認為是與有機廢物汙染和不應該用於灌溉使用。

31. B. 馬桶沖洗

   D. 滴灌

   黑水不應該接觸人類。噴灌技術有可能噴灑到人或寵物，所以，最好使用滴灌技術來使用，因為滴灌灌溉水分佈在土地裡面，而不是在它上面。

32. A. 玻璃

   D. 塑膠

   E. 紙板

   雖然 LEED 鼓勵所有回收，但是專案只需提供回收利用的玻璃、塑膠、金屬、紙和紙板。

33. C. 地毯

34. D. ASHRAE 標準 90.1-2012 年

其他列的 ASHRAE 標準涉及其他問題，如通風與熱舒適。BIFMA 是
美國辦公傢具製造商協會（The Business and Institutional furniture Manufacturer's Association）的英文簡稱。美國辦公傢具製造商協會是一個
非盈利性組織，作爲信息資源中心和辦公傢具工業代表，BIFMA 爲
北美洲辦公傢具製造工業提供專業且廣泛的貿易發展機會。BIFMA
針對辦公傢具穩定性、強度及疲勞性等性能制定標準，其標準嚴格、
完善，贏得了廣泛的認同。

35. A. 認證木材

森林管理委員會（FSC）認證，認證之木材的認證資料和供應鏈文檔。

36. A. LEED 商業內裝
    D. LEED 營運和維護

核心和外殼的 LEED 是專門處理有關由不同的內部承租客空間的建築
認證。住宅版則是針對住宅的認證，新建築或學校還爲不同的產品範
圍和種類的新建築物認證，營運和維護則可以在建築物營運後取得。

37. C. 氫氟碳化合物（HFCs）

38. A. LEED 營運和維護

LEED O＋M 是目前唯一評分系統要求換發新證的，至少每五年需要
重新認證一次。

39. B. 地熱交換系統

　　E. 核電

40. A. 大學學生

　　E. 零售商店客戶

　　這些瞬間的遊客預計在建築物中將只花一個小時左右，而不是整天，而小學學生或辦公室工作人員則是一整天。

41. E. 以上皆非

　　生命週期成本法考慮到所有的清單所列因素。

42. E. LEED 住宅

43. A. 認證會被撤銷

44. B. 本地物種

　　C. 適應物種

　　本地和適應的植物材料都能夠在沒有太多人照料的情況下生存，例如灌溉和原來的氣候下生長。另一方面，栽培的品種往往需要大量的維護和澆水。外來入侵物種可能需要或不需要灌溉，但他們可以有其他的負面影響，例如扼殺了生物多樣性，所以應該避免。

45. B. 為連結到網站 www.usgbc.org/leed

46. B. 廁所

　　廁所的廢水被認為是黑水，也不能拿來使用。

47. A. 土地清理的廢土

　　C. 石棉地磚

　　應該從建築廢物管理計算中排除清理土地的碎片和石棉等有害物質。

48. E. HFC 23

49. D.針葉樹的樹苗下停車

雖然這些樹最終可能為停車場提供大量的樹蔭，但是目前這些微小的樹木沒有提供任何重大的樹蔭，LEED 提到只能算在專案完成的五年內產生之樹木。

50. A.大廈本身

E. 停車很多

F. 人行道上

51. D.使用瀝青停車地區

典型的瀝青是非常黑暗的表面，會增加熱島效應，而不是最小化。

52. E. 快速可再生材料和區域材料

棉花是少於十年可持續增長和收穫的農業產品，因此，有資格作為一種快速可再生材料。既然產品也種植並在距離專案 100 英里範圍內製造，它也有資格作為本地區域材料。

53. C. 可再生能源證書

54. C. 他們都不能計算得分

在建築設計和施工評分系統，一種材料必須被提取、收穫和加工或製造要考慮區域的材料，超過 100 英里範圍內是沒有辦法得分的。

55. D.LEED 住宅

56. E. 反射率

這兩個術語是指表面反射的光量。

57. B. 使用屋頂材料具有高的太陽能反射

E. 使用植被屋頂

58. C. 給予專案得分

雖然 LEED 專案管理員將每個得分項目作記錄，但是得分並非由 LEED 專案負責人決定。

59. A. 透過當地水電公司提供之方案

B. 透過購買綠色電子認證

「綠色能源」是異地，網格電網來自可再生能源。現場太陽能安裝量將可以被視為「現場可再生能源」，但是不是「綠色能源」。來自當地風力發電的能量，如果沒有綠色電力證書也無法得到分數。

60. C. 本地植物

因為綠色屋頂可能難以提供一般使用，他們種植時應該是低維護，其中許多本地物種很適合，而大多數一年生草本和栽培物種的材料需要灌溉和施肥。樹木和灌木可能有根生長很深和可能自重太高或是破壞結構。

61. A. 空氣品質

SMACNA 主要是制定施工期間的建築物的室內空氣品質。

62. B. 用雨水沖洗馬桶

C. 最小化停車空間

63. D. 褐地

E. 以前開發過的基地

LEED 評分系統避免在農田、生態為瀕危物種或接近水體的基地上進行開發。相反，評分系統希望建築於褐地或以前開發過的土地（發展密度和社會連接）。

64. B.必須符合經濟效益

正因為「創新的」創新與設計的得分的許多策略，會非常昂貴。

65. D.優先停車位必須分布在所有地區

66. D.臭氧層破壞

67. A.草木茂密

開放空間必須茂盛，但它並沒有規定植被是本地或適應的物種，也就是說任何植被是可以接受的。

68. A.氟氯碳化物

氟氯碳化物的冷媒在一般情況下，是最有效率的冷媒。但已經逐步淘汰，因為它們有負面的環境影響。

69. C.它可能比分類回收的成本更高

很多廢物管理公司已在網站上顯示，混合收取比分類收取的費用更貴，因為還需要額外的人工來做分類處理。

70. A.距離

C.高速公路交匯處

D.牆

你可以經常簡單地步行通過或到達停車場或公園，但牆或高速公路的交匯處可能創建不可逾越的障礙。因為人們願意走的距離有限，距離也會妨礙行人通道。

71. D.ADA（美國殘疾人法）

民權法案也是一項聯邦法，但它與建築法規無關，能源政策法案是一項聯邦法與相關的建築規範，但它涉及能源和用水、不管理無障礙。ASHRAE 是能源規範，但不屬於聯邦法案。ANSI 是全國研究所的設置標準。

72. D. 喝

73. A. 噴霧灌溉

噴灌技術使用的水量滴灌等更多。

74. B. 源頭減量

75. A. 在專案前期策劃階段

綜合設計過程應盡可能早在過程開始。

76. A. 遊船上的賭場

D. 臨時體育設施

LEED 最低程式要求專案必須在現有土地上興建建築物，以及建築必須是永久的。這兩項要求讓上述的專案沒辦法獲得 LEED 認證。

77. A. 室內熱水器

78. B. 廚房水龍頭

C. 淋浴蓮蓬頭

D. 浴室水龍頭

馬桶和小便池之用水量的單位是每沖洗一次幾加侖（GPF）

79. C. 能源和大氣

80. A. 獲得卓越性能

卓越的性能可以透過超過得分的基準值得到。

81. C. 建立建築幫浦的設備性能基準

能源政策法案也處理能源使用，但是 LEED 使用它來建立基線水使用費率衛生設備。

82. C. 生命週期成本

生命週期成本計算考慮到不僅是產品和系統的第一次費用，也是日常

維護、能源使用、壽命等的費用。可持續設計策略往往具有更高的初始成本，但是可以顯示是在長期內由於減少能源或維修費用而整體更便宜。

83. D.以前開發的基地

LEED 鼓勵選用開發過的土地。

84. A.美國加州空氣資源局 ZEV

B.美國經濟節能委員會的 Green Book

85. A.1

一個專案只可以得到一分，在創新和設計的部分無論多少位 LEED AP 參與都只能得到一分。

86. D.50%

87. D.專案得分表格

一些得分和系統必備元件可能還需要進一步的佐證，但所有 LEED 得分和系統必要條件都必須要求表格。

88. A.專案面積

89. E. 木材

雖然木頭是農業產品，它的生長和收穫週期超過 10 年。

90. D.100 英里

91. B.最低專案要求

D.系統必備元件

92. D.計程車

為滿足 LEED 要求必須為「軌道交通」，且必須乘載至少十位或以上的人，而且有固定的時程表。

93. C. LEED 新建建築版

高層住宅專案（4 層或更多）必須為新建建築，而不能使用 LEED 住宅版。

94. A. 侵蝕

E. 沉澱

F. 空氣中粉塵汙染

95. B. 黑水

因為馬桶產生出的水包含有機廢物，它被認為是黑水，也稱為下水。

96. B. 1.0 GPF

由於乾燥的小便池沒有水，因此可以減少一般小便斗 1.0 GPF 的用水量。

97. B. 10

這些十個得分表示從創新和設計與區域優先類別取得。

98. E. 空間

99. C. 室內空氣品質

雖然室內空氣品質是重要的類別，類別的名稱是室內環境品質，不單單只包含空氣。

100. B. 用水高效灌溉

第十二章 附 錄

12.1 LEED 考試常見問題 Q&A

LEED AP 和 LEED GA 考試八問八答

結合考試要求，稍微總結了一下常見問題解答。

問 1：我是一個 LEED GA，我想成爲 LEED AP，要參加什麼考試？

答： 如果你是一個現行的 LEED GA（現行的意思是參加證書維護更新），只要再參加專業的 LEED AP 考試通過即爲 LEED AP，同時 LEED GA 證書維護也將被 LEED AP 代替。

問 2：需要在參加 LEED GA 後才有資格參加 LEED AP 嗎？

答： 不需要，LEED GA 是 LEED AP 考試的第一部分，因此不需要。

問 3：如果我只通過了 LEED AP 考試的一部分？

答： 如果只通過了 LEED AP 考試的一部分，必須重新參加未過的那部分考試。也就是說，在申請時限內，三次的考試機會還剩兩次。

問 4：如果我過了 LEED GA ，但沒過 LEED AP 專業考試，能稱自己是 LEED GA 嗎？

答： 不行。因爲你申請的是 LEED AP 考試，你有三次機會過 LEED GA+LEED AP 專業兩部分考試，而不是其中一部分的 GA 考試。

問 5：如果我通過了第二部分 LEED AP 專業考試，但未過第一部分 LEED GA 考試，能稱自己是 LEED AP 嗎？

答： 不行。若想獲得 LEED AP，必須要兩部分一起通過。正如前面所說，只要重新參加未過的那部分考試。也就是說在申請時限內，三

次的考試機會還剩兩次。

問 6：在我的申請時限內，三次未能過 LEED GA 考試，怎麼辦？

答：　如果是這樣，那已經用完了三次機會。必須要等到申請期限到期（一年到期），順延 90 天後才能重新申請。

問 7：如果已經過 LEED GA 考試，但是 LEED AP 考試連考三次沒過，怎麼辦？

答：　要等待第二次的申請，或者選擇另一個專業考。可以申請 LEED GA 的證書維護計劃，如果不申請，那麼請重新申請 LEED AP 考試。

問 8：我能同時成為 LEED GA 和 LEED AP 嗎？

答：　不行。只能是 LEED GA 或者是 LEED AP。如果你已經成為 LEED AP（考試通過或登記註冊），就不再使用 LEED GA 名稱，同時 LEED GA 證書維護也將被 LEED AP 代替。

12.2 英文名詞解釋表（English Glossary）

Term (English)	繁體中文名詞	中文名詞定義
abandoned property	廢棄物業	有意和永久遺留下的財產，且之前的業主似乎並不準備回來認領或使用之。未履行合同要求可視為放棄合同財產（物業）的權益。但是，地役權和其他土地權益不能只因為未使用而視為廢棄財產（物業）。廢棄土地被定義為目前未使用的土地，但可能有現成的公用設施和基礎設施。
adapted plant	適應性植物	不是特定區域原生的植物，但擁有可以讓其在該區域生存的特性。適應性植物不會帶來與入侵物種一樣的問題。

Term (English)	繁體中文名詞	中文名詞定義
added antimicrobial treatment	抗菌處理添加劑	添加到產品（如，油漆、地板）中以殺死或抑制微生物生長的物質。某些產品，如油氈，具有天然的抗菌特性。儘管目前有所實踐，科學尚未證明抗菌處理能比標準清潔程式更有效地減少建築飾面中細菌的感染傳播。也被稱爲微生物添加劑。參見美國環保局事實表《經殺蟲劑處理的消費類產品》（www.epa.gov/pesticides fact-sheets/treatart.htm）。
adjacent site	相連基地	至少有 25% 的連續邊界與先前已開發基地地塊接壤的基地。僅考慮不干擾道路用地的接壤地塊。計算中不考慮任何與水體接壤的邊界部分。
alternative daily cover (ADC)	每日覆蓋替代材料（ADC）	土質材料除外的一種材料，每個運營日結束時，放在城市固體廢物填埋場作用面的表面，用於控制病菌、火災、氣味，防止垃圾被吹飛，以及防止人員撿拾垃圾堆裡的東西。一般來說這些材料必須經過處理，這樣才不會在裸露的垃圾填埋場表面留下空隙。（CalRecycle）
alternative fuel	替代燃料	低污染非汽油燃料，如電力、氫氣、丙烷、壓縮天然氣、液化天然氣、甲醇及乙醇
alternative water source	替代性水資源	除來自公共設施、基地內地表水源和地下天然淡水水源以外的非飲用水，包括灰水、基地內再生水、收集的雨水、收集的冷凝水，以及反滲透系統（IgCC）的作業廢水。
annual sunlight exposure (ASE)	年陽光照射度（ASE）	一種描述室內工作環境潛在視覺不適程度的指標。被定義爲分析區域中超出指定陽光直射照明等級的時間超過指定小時數的面積百分比。

Term (English)	繁體中文名詞	中文名詞定義
appurtenance	附屬構件	屋面系統內置的非結構部分。包括天窗、通風裝置、機械設備、隔斷和太陽能板。
ASE1000,250	ASE1000,250	使用最大 2 英尺的點間距，考慮將相同的 10 小時／天分析期作爲 sDA，並使用可比的類比方法，報告分析區域中在安裝活動窗簾或遮陽設備以阻擋陽光之前，每年暴露在超過 1000 勒克斯直射陽光下超過 250 小時的感測器的百分比
assembly	組合材料	由多種材料組成的產品（例如，混凝土），或者由子部件組成的產品（如工作站）
attendance boundary	學生活動邊界	教育局用來根據學生居住的地方來確定學生上哪所學校的邊界
average LED intensity (ALI)	平均 LED 密度（ALI）	發光二極體燈的照明輸出，由國際照明委員會標準 127-2007 指定
base building	基本建築	組成建築的，或者永久性和半永久性安裝在專案中的材料和產品（如地板、箱櫃、牆面）
baseline build-ing perfor-mance	建築性能基線	建築設計的年能源成本，用於與高於標準要求的設計進行比較的基線
baseline condi-tion	基地基本狀況	在 LEED 專案開始之前，但是不一定要在開發或施工開始之前。基地基本狀況描述了在開發商通過購買或購買選擇權獲得大部分可建土地的權利那天專案基地的狀態。
baseline water consumption	用水量基線	與專案設計或實際水錶數相比，在採用符合規範且此外不會節省更多用水的設備及配件時，對建築用水量的計算預測

Term (English)	繁體中文名詞	中文名詞定義
basis of design (BOD)	設計基礎要求（BOD）	完成業主專案任務書所需的資訊，包括系統介紹、室內環境品質條件、設計假設和對適用條例、標準、規範和指南的參考
bicycle network	自行車道網路	由以下任意組合構成的連續網路 1) 街道外的自行車道或小徑，雙車道至少 8 英尺（2.5 米）寬，單車道至少 5 英尺（1.5 米）寬 2) 在街道上實際指定的至少 5 英尺（1.5 米）寬的自行車道 3) 設計速度為 25 mph（40 km/h）的街道
bicycling distance	自行車騎行距離	騎自行車的人必須在起點和終點之間騎行的距離，整個騎行過程都必須在自行車道網路中進行。
bio-based material	生物基材料	完全或大部分由生物製品、可再生農業材料（包括植物、動物和海洋生物）或林業材料構成的商業或工業產品（食品或飼料除外）。就 LEED 而言，這不包括皮革和其他動物皮。
blackwater	污水	含有尿或糞便，且應被排放到符合國際管道規範的建築或住房衛生排水系統的廢水。根據某些州或地方法規，廚房水槽（有時會根據垃圾處理的使用情況進行區分）、淋浴或浴缸廢水被視為污水。
blowdown	排放	從冷卻塔或蒸發式冷凝器循環系統中去除補給水以降低溶解固體的濃度
brownfield	褐地	因含有或可能含有危害性物質、污染物或致汙物而使得擴大、再開發或再利用變得複雜的不動產。
BUG rating	BUG 評級	一種光源分級系統，按照背光源 (B)、向上照射的燈 (U) 和眩光 (G) 對光源進行分級（依據 IES/IDA 照明條例範本）。BUG 評級取代之前的截光評級。

Term (English)	繁體中文名詞	中文名詞定義
buildable land	可建造土地	可以進行建造的基地部分，包括自願預留和未建造的土地。在密度計算中使用時，可建造土地不包括公共道路用地和現行法律禁止開發的土地。
building exterior	建築外構造	建築的主要和輔助防水系統，包括防水膜以及密封和防水阻隔材料，以及該系統外的所有建築構件
building interior	建築室內	建築防水／防潮層以內的所有構件
bus rapid transit	公車快速／專用通道	經過改善的公車系統，只在公車專用車道或其他公交道路用地上運營。該系統旨在將公車的靈活性與軌道的效率相結合。
carbon offset	碳補償	爲補償其他地點的碳排放而減少、避免或隔離的二氧化碳當量單位（世界資源研究所）
chain of custody (CoC)	監管鏈（CoC）	用於跟蹤產品的程式，從收穫或萃取到其最終用途，包括加工、轉化、製造及分銷這幾個連續的階段
charrette	專家研討會	深入細緻的多方研討會，將不同學科和背景的人聚集到一起來探索、形成和共同制定設計方案
chlorofluorocarbon (CFC)-based refrigerant	氯氟化碳（CFC）型冷媒	一種含有氯氟化碳的液體，在低溫條件下從儲液器中吸熱並在高溫條件下散熱。排放到大氣中以後，CFC 會消耗平流層的臭氧層。
civil twilight	民用曙暮光	早上（黎明）或晚上（黃昏）太陽中心線在地平線以下 6 度的時間。在天氣良好的情況下，民用曙暮光是明確區分地面物體的最佳時間。早上在民用曙暮光之前，以及晚上在民用曙暮光之後，一般來說都需要人工照明以進行普通戶外活動。

Term (English)	繁體中文名詞	中文名詞定義
classroom or core learning space	教室或核心學習空間	定期占用並用於教育活動的空間。在這種空間裡，主要功能就是教學，良好的語言溝通對學生的學術成就至關重要。（引自 NSI S12.60）
clean waste	清潔廢棄物	建造與拆除後遺留的非有害物質。清潔廢棄物不包括鉛和石棉。
clear glazing	透明玻璃	透明且用作窗戶時視野良好的玻璃。散射玻璃只能進行自然採光。
closed-loop cooling	閉環冷卻	使用迴圈水並作為排熱建築和醫療設備散熱器的系統。由於水密封在系統中，某些閉環冷卻系統使用非飲用水（如從空氣處理裝置的冷卻盤管冷凝器回收的工藝用水）。
color rendering index	顯色指數	一種從 0 到 100 的指標，表示人造光源相比白熾燈在顯示色調方面的準確性。指數越大，燈呈現的顏色就越準確。白熾燈照明的顯色指數高於 95；標準高壓鈉燈照明（如橙色色調的巷道燈）指數大約為 25；很多使用稀土螢光粉的螢光燈光源的顯色指數為 80 或更高。（引自美國能源之星（ENERGY STAR））
combination oven discharge	組合爐排水	包括蒸汽迴圈或選件的爐子中排出的水
combined heat and power	熱電聯合	一種綜合系統，收集發電過程中未使用的、由單一燃料來源產生的熱量。也被稱為熱電聯產。（引自美國環保局（EPA））
commingled waste	混合廢棄物	建築廢棄物流在專案基地中混合並運走，之後再分揀成可回收的廢物流。也被稱為單流回收。

Term (English)	繁體中文名詞	中文名詞定義
commissioning (Cx)	調試（Cx）	檢驗與記錄一幢建築及其所有系統與元件如何在規劃、設計、安裝、測試、操作及維護上滿足業主專案要求的過程
commission-ing authority (CxA)	調試機構（CxA）	負責組織、領導和審查調試過程活動是否完成的機構。CxA 有助於業主、設計人員和承建商之間的溝通，可以確保根據業主專案任務書安裝和運行複雜系統。
conductivity	導電率	利用電流在水中傳導的能力，對水中溶解固體的濃度水準進行測量的指標。由於其受溫度的影響，爲了標準起見，將在 25°C 時測量導電率。
conventional irrigation	常規澆灌	某地區最常見的通過非自然方式爲植物供水的系統。傳統的灌溉系統通常使用壓力供水，並通過灑水噴頭澆灑地面。
cooling tower blowdown	冷卻塔排放	從冷卻塔中排放水，一般是由於鹽度或鹼度越來越高，導致結垢。冷卻塔排放的水鹽度過高，無法在景觀灌溉中使用。
cradle-to-gate assessment	搖籃到大門評估	對產品部分生命週期的評估，從資源採集（搖籃）到工廠大門（發運經銷和銷售之前）。其省略了產品的使用和廢棄處理階段。
cultural land-scape	文化景觀	一個官方指定的地理區域，包括與歷史事件、活動或人物相關，或者展現出其他重要的文化或審美價值的文化和自然資源
current fa-cilities require-ments (CFR)	現有設備要求（CFR）	業主專案要求的實施，經過制定以確認業主的當前運行需求和要求

Term (English)	繁體中文名詞	中文名詞定義
dedicated storage	專用存儲區	建築空間或中央設施中的一個專用區域，針對具體任務（如收集可回收廢棄物）確定其大小並加以分配。標牌往往指示存放在這裡的可回收廢棄物的類型。某些廢棄物流，如含汞燈、敏感的紙質檔、生物醫療廢棄物或電池，可能需要特別的操作和處理方法。請參考市政當局的安全存放和處理常式，或使用美國環保局網站 www.epa.gov 上張貼的指南。
demand response (DR)	需求回應（DR）	需求側資源為回應電價變化或設計用於在市場批發價較高或者系統可靠性受影響時降低用電量的獎勵而作出的與正常消耗模式不同的用電量變更
demand response (DR) event	需求回應（DR）事件	公用設施或獨立服務提供者呼籲計畫參與者改變電網電力使用模式或使用量的一段指定時間。也被稱為限電事件。
demountable partition	靈活隔斷	可以輕鬆重新配置的臨時內牆。在注重聲學效果的擁有嵌入式設備的衛生保健設施中（如手術套房），可能會禁止使用靈活隔斷
densely occupied space	人員密集使用空間	設計使用密度為每 1000 平方英尺 25 人或以上的區域（人均 93 平方米）
density	密度	地塊上的建築總面積或居住單元相對於該地塊可建造土地面積的比值。測量密度的單位會因得分點要求而有所不同。不包括結構性停車場。
departmental gross area (DGA)	部門總面積（DGA）	一 臨床部門診斷和治療設施的建築面積，從分隔部門與相鄰空間的牆壁中心線開始計算。部門內的牆壁和交通空間將包括在計算中。該計算不包括住院病房。

Term (English)	繁體中文名詞	中文名詞定義
development footprint	開發占地	專案基地（包括作為專案一部分開發的建築、街道、停車場以及其他常規不滲透的表面）的總占地面積
differential durability	差異耐久性	使用壽命不同的兩種材料組成一個完整元件的一種狀態。如果一種材料磨損且無法分離或更換，整個產品必須拋棄。
direct access	直接入口	一種進入空間的方法，無需離開當前樓面或穿過其他患者的房間、員工專用空間、服務或公用設施空間，或者主要的公共空間。患者和公共交通走廊、共同的休息區以及等待和日間區域都可以是直接入口路線的一部分。
direct sunlight	直射日光	室內測得 1,000 勒克斯或更多的直射日光量，計入窗戶透光率和角度影響，不包括任何活動窗簾的影響，也不包括反射光（即，零反射分析）和散射天空元素的影響（引自 IES）
district energy system (DES)	區域能源系統 (DES)	一種中央能源轉換站以及傳送和分配系統，為一組建築提供熱能（如，大學校園中的中央供冷站）。不包括僅提供電力的中央能源系統。
diverse use	多樣化使用	提供旨在滿足日常所需的商品或服務的不同企業或組織，公開可用。不包括 ATM、或自動販賣機等自動化設施。如需完整列表，請參閱附錄
downstream equipment	下游設備	專案建築或專案基地中的加熱和冷卻系統、設備和控制裝置，與將區域能源系統（DES）的熱能輸送到加熱和製冷空間有關。下游設備包括與 DES 之間的熱連接或介面、建築中的輔助分配系統和終端設備、由穿過系統的氣流從冷卻塔或蒸發式冷凝器帶來的漂流水滴。

Term (English)	繁體中文名詞	中文名詞定義
durable goods	耐久產品	使用壽命大約爲 2 到 3 年且不頻繁更換的產品。包括傢俱、辦公設備、家用電器、外部電源適配器、電視機，以及視聽設備。
durable goods waste stream	耐久產品廢棄流	來自專案建築的已完全折舊且已達到正常業務運營使用壽命的耐久產品流。包括出租並還給所有者的耐久產品，但不包括仍可使用且轉移到其他樓層或建築的耐久產品。
electric vehicle supply equip-ment	電動車供電設備	包括不接地、接地和設備接地導體在內的導體、電動車連接器、連接插頭和其他所有配件、設備、電源插座或專門爲了從樓宇電纜向電動車提供電力的用具。（國家電氣規範和加利福尼亞州 625 條款）
electronic waste	廢棄電器	丟棄的辦公設備（電腦、顯示器、影印機、印表機、掃描器、傳眞機）、家用電器（冰箱、洗碗機、飲水機）、外部電源適配器、電視和其他視聽設備
elemental mer-cury	純汞	最純形態的汞（而不是含汞化合物），其蒸氣通常用於螢光燈和其他類型的燈泡
emergency lighting	應急照明	僅在緊急情況下工作的光源，在常規建築運營期間始終關閉
employment center	就業中心	至少 5 英畝（2 公頃）的非居住區，就業密度至少爲每淨英畝 50 名員工（每淨公頃至少 125 名員工）
enclosure	圍護結構	建築外部和半外部結構。外部包括將有空調的空間與外部相隔離的建築元素（例如，牆體元件）。半外部包括將有空調的空間與無空調的空間相隔離的建築元素，或者封閉半加熱空間的建築元素，熱能可以通過該空間與外部、有空調的空間或無空調的空間（如，閣樓、爬行空隙、地下室）進行傳遞。

Term (English)	繁體中文名詞	中文名詞定義
energy service provider	能源服務供應商	可以讓外部實體，如 USGBC 訪問建築管理團隊使用能源之星資料管理器或類似工具維護的水和能源使用資訊的機構
engineered nanomaterial	工程納米材料	在分子（納米）水準上設計的物質。由於尺寸很小，因此具有其大部分傳統競爭對手一般不具有的新穎特性。參見澳大利亞國家工業化學品通報和評估計畫，nicnas.gov.au/publications/information_sheets/general_information_sheets/nis_nanomaterials_pdf.pdf。
environmental product declaration	產品環境聲明	聲明該物品符合 ISO 14021-1999、ISO 14025-2006 和 EN 15804 或 ISO 21930-2007 的環保要求
evapotranspiration	蒸散	進入大氣的蒸發和植物蒸騰之和。蒸發是指土壤、植物表面或水體的液體變為蒸氣。蒸騰是指水通過植物轉移以及隨後以水蒸氣形式散失。
extended producer responsibility	生產商延展責任	產品製造商採取的措施，以許諾其自己的、有時還包括其他製造商的產品在產品使用壽命結束後將成為消費後廢棄物。生產商將回收材料以用於相同類型的新產品。為了符合得分點要求，必須廣泛採用某種方案。對於地毯，擴大生產商責任必須與 NSF/ANSI 140-2007 保持一致。也被稱為閉環程式或產品收回。
extensive vegetated roof	粗放型種植屋面	覆滿植物且一般並不被設計為可隨便出入的屋面。通常，粗放型系統是堅固耐用的種植屋面，建造好之後幾乎不需維護。粗放型種植屋面中的種植介質厚度從 1 到 6 英寸。（引自美國環保局（EPA））外殼種植表面面積，專案基地中植被總面積，包括種植屋面和草皮

Term (English)	繁體中文名詞	中文名詞定義
external meter	外部水錶	安裝在水管外部以記錄通過的水量的設備。也被稱爲鉗型表。
floor-area ratio (FAR)	容積率（FAR）	非居住用地的密度，不包括停車場，計算方法爲非居住建築總面積除以用於非居住結構的可建用地總面積。例如，在有 10000 平方英尺（930 平方米）可建用地面積的基地內，FAR 爲 1.0 則表示有 10000 平方英尺（930 平方米）的建築面積。在相同的基地中，FAR 爲 1.5 則表示有 15000 平方英尺（1395 平方米），FAR 爲 2.0 則表示有 20000 平方英尺（1860 平方米），FAR 爲 0.5 則表示有 5000 平方英尺（465 平方米）的建築面積。
foundation drain	基礎排水	從地下排水系統排出的水。如果建築地基低於地下水位元，則可能需要使用集水坑泵。從水坑排出的水可以儲存起來並用於灌漑。
freight village	物流園	一群包括聯運轉移業務的貨運相關企業。物流園可提供物流服務、集成配送、倉儲能力、展示廳和支援服務。這些支援服務可包括安保、維護、郵遞、銀行、海關和進口管理協助、食堂、餐廳、辦公場所、會議室、酒店、公共或活動中心運輸。
functional entry	功能入口	設計由行人使用且在正常營業時間開放的建築開口。不包括任何專門設計作爲緊急出口的門，或者並沒有被設計爲行人入口的車庫門。
furniture and furnishings	傢俱和陳設	爲專案購買的獨立傢俱，包括獨立和多人座椅；開放式和私人辦公室工位；書桌和桌子；存儲裝置、餐具櫥、書架、檔案櫃和其他箱櫃；壁裝式視覺顯示產品（如，標誌板和佈告板，不包括電子顯示器）；以及其他陳設，如

Term (English)	繁體中文名詞	中文名詞定義
		畫架、手推車、獨立展架、安裝的織物和靈活隔斷。根據專案需要，還可包括酒店傢俱。不包括辦公配件，如桌面吸墨紙、託盤、磁帶分配器、廢物籃和所有電器產品，例如照明設備和小器具。
grams per brake horse-power hour	每制動馬力小時克數	用於描述具有特定馬力的引擎在一小時內排放的排放物（如，氮氧化物或顆粒物質）克數指標
graywater	灰水	「尚未接觸廁衛垃圾的未經處理的生活廢水。灰水包括浴缸、淋浴間與浴室面盆洗滌用水以及洗衣機與洗衣池用水，但不包括廚房水槽或洗碗機廢水。」（統一給排水規範附錄 G，獨戶住宅灰水系統）；「衛生間、浴缸、淋浴間、洗衣機和洗衣水槽排放的廢水」（國際給排水規範附錄 C，灰水回收系統）。某些州和地方當局還將廚房水槽廢水納入灰水範疇。州與地方法規中可能存在其它差異。專案團隊應遵循專案所在地區的管轄當局確定的灰水定義。
green infra-structure	綠色基礎設施	一種基於土壤和植被的潮濕天氣管理方法，具有成本效益、永續性和環保性。綠色基礎設施管理方法和技術包括滲透、蒸騰、收集和重複利用雨水，以保持或恢復自然水文狀況。（引自美國環保局（EPA））
green power	綠色電力	一部分由可再生能量資源產生的電網電力組成的可再生能源
green vehicles	綠色汽車	在美國節能經濟委員會（ACEEE）的年度車輛等級指南（對於美國以外的專案，可以是當地的同等標準）中至少能得到 45 分綠色分數的車輛。

Term (English)	繁體中文名詞	中文名詞定義
greenfield	未開發荒地	之前未開發、分級或受到干擾，並且可作空地、棲息地或自然水文用途的區域
hardscape	硬景觀	建築景觀中無生命的元素。包括鋪過的路面、道路、石牆、木材和合成地板、混凝土路徑和人行道、混凝土、磚，以及瓷磚鋪砌的庭院。
hazardous material	有害物質	自身或通過與其他因素相互作用有可能對人類、動物或環境有害的任何物品或媒介（生物、化學、物理）
heat island effect	熱島效應	硬景觀（如黑色非反射路面與建築）對熱量的吸收及其對周邊地區的輻射。其他促成因素還可能包括汽車尾氣、空調和街道設備。高層建築與狹窄街道減少了氣流並加劇了這一效應。
highway	高速公路	用於汽車的運輸專用通道，只有有限的出入口，禁止人力車，比地方道路速度更快。高速公路一般連接城市與城鎮。
historic building	歷史建築	具有歷史、建築、工程、考古或文化意義的建築或結構，列為或被認為有資格作為歷史結構或建築，或者在指定的歷史街區中作為有影響的建築或結構。"歷史"稱謂必須由當地歷史保護審查委員會或類似機構給出，結構必須列在州史跡名錄、國家史跡名錄（在美國以外符合當地同等標準），或者被認為有資格列在上述名錄中。
historic district	歷史街區	被認為或者確定有資格具有歷史和建築重要意義的一組建築、結構、物體和基地，無論對該地區的歷史本質是否有影響

Term (English)	繁體中文名詞	中文名詞定義
homogeneous material	同質材料	只包括一種材料或者包括無法以機械方式分離的多種材料組合的物品，表面塗層除外
hydrozone	水電區	具有類似水需求的一組植物
illuminance	照度	位於某點且朝向特定方向的表面差異元素上的入射光通量密度，以每單位面積的流明來表示。由於涉及區域有差異，習慣性地將其稱為某點的照度。單位名稱取決於區域測量單位：如果使用平方英尺作為面積單位，則為尺燭光；如果使用平方米，則為勒克斯。（引自 IES）在專業術語中，照度是對投射到表面的光量的測量。其在美國表示為尺燭光（基於平方英尺），在大多數其他國家表示為勒克斯（基於平方米）。
impervious surface	非滲透表面	經開發和建築改造過的一塊區域，致使降水無法通過土壤向下滲透。非滲透表面的例子包括屋面、鋪設的道路和停車場、人行道以及因設計或使用而變得堅實的土壤。
individual occupant space	個人使用空間	住戶執行各自不同任務的區域。個人使用空間可能位於多住戶空間中，應盡可能區別對待。
industrial process water	工業工藝用水	從工廠排出的任何水。在這種水可用於灌溉之前，需要檢查其品質。鹽水或腐蝕性的水不得用於灌溉。
infill site	嵌入式場地	專案邊界 1/2 英里（800 米）之內至少 75% 的土地面積（不包括道路用地）是之前已被開發過的基地。街道或其他道路用地不構成之前已開發的土地；重要的是道路用地或街道另一側物業的狀態。

Term (English)	繁體中文名詞	中文名詞定義
infiltration	滲入	（HVAC）空氣從無空調的空間或室外因壓力差（此壓力差同樣會導致滲出）而不受控制地通過天花板、地板和牆壁上的非人為開口向內洩漏到有空調的空間。（ASHRAE 62.1-2010）
infrared (thermal) emittance	紅外線（或熱）發射度	介於 0 和 1（或 0% 和 100%）之間的一個值，表示一種材料散發紅外線輻射（熱量）的能力。涼爽的屋面應該具有較高的熱發射度。輻射能量的波長範圍大體上是 5 至 40 微米。大多數建築材料（包括玻璃）在這個光譜區段內是不透明的，發射度大約為 0.9，或 90%。乾淨的裸金屬，如未失去光澤的鍍鋅鋼，發射度較低，是發射度不為 0.9 的最主要情況。相比之下，鋁屋面塗層具有中等發射度。（引自 Lawrence Berkeley 國家試驗室）
inpatient	住院病人	住到醫療、外科、產科、專業或重症監護病房並停留超過 23 小時的人
inpatient unit	住院病房	任何讓患者接受護理超過 23 小時的醫療、外科、產科、專業或重症監護病房
integral labeling	內置標籤	無法輕易刪除的資訊傳達系統。對於傢俱來說，這種標籤可包括有關材料產地、屬性和生產日期的射頻識別、雕刻、壓花，或其他包含資訊的永久性標記。
integrated pest management	綜合害蟲管理	一種蟲害治理方法，可保護人類健康和周圍環境，並通過最有效、風險最低的方案提高經濟效益
integrated project delivery	整合式工程交付	一種涉及人、系統和業務結構（合同與法律協定）與實踐的方法。該過程利用所有參與者的才能和見解，在設計、製造和建造的所有階段改善結果、為業主增加價值、減少浪費並最大程度提高效率。（引自美國建築研究院）

Term (English)	繁體中文名詞	中文名詞定義
intensive vegetated roof	密集種植屋面	這種屋面相比粗放型種植屋面，有更多的土壤量，可種植更多種類的植物（包括灌木和樹），並實現更多的用途（包括人類使用）。生長介質的厚度是決定棲息地價值的一個重要因素。為屋面選擇的本地或可適應性植物應能滿足基地特有野生動物種群的需求。（引自 Green Roofs for Healthy Cities）
interior floor finish	室內樓地板面層	完工的底層地板或樓梯（包括樓梯踏板和豎板、斜坡和其他行走表面）上使用的所有面層。室內面層不包括建築結構構件，如梁、桁架、螺柱或底層地板或類似物品，也不包括不完全擴散的濕塗料或粘合劑。
interior wall and ceiling finish	內牆和吊頂面層	構成建築裸露室內表面的所有面層，包括固定牆壁、固定隔斷、柱子、裸露的吊頂和室內護牆板、鑲板、內飾或者通過機械方式安裝或用於裝飾、聲學校正、表面耐火或類似目的的其他飾面
intermodal facility	聯合運輸設施	先後使用兩種或更多運輸模式來移動單個裝載裝置或道路車輛中的貨物，而無需處理貨物自身的設施
invasive plant	入侵植物	被引進到某地區且有侵略性地適應和繁殖的非本地植物。植物的活力加上缺少天敵往往會導致數量激增。（引自美國農業部）
IT annual energy	IT 年能耗	資訊技術和電信設備（包括伺服器、網路和存放裝置）的年耗電量
lamp	燈	在燈具中發光的設備，不包括燈罩和鎮流器。以發光二極體作為光源的傳統燈也符合此定義。

Term (English)	繁體中文名詞	中文名詞定義
lamp life	燈壽命	人造光光源（如燈泡）的有效工作壽命。螢光燈使用壽命的確定方法是：每關閉 20 分鐘後打開三小時。對於高密度放電燈，測試方法為每關閉 20 分鐘後打開 11 小時。燈壽命取決於啓動鎭流器是程式式的還是暫態的。這些資訊公佈在製造商資訊中。也被稱爲額定平均壽命。
land trust	土地信託	一種私有的非營利組織，其全部或部分使命是，通過從事或幫助保護地役權或土地收購，或者管理這些土地或地役權，積極致力於節約土地（引自土地信託聯盟）
land-clearing debris and soil	土地清理碎片和土壤	天然材料（如，岩石、土壤、石頭、植被）。即使在基地中，人造材料（如，混凝土、磚塊、水泥）也會被認爲是營建廢棄物。
landscape water requirement (LWR)	景觀用水要求（LWR）	基地景觀區域在基地峰值用水月份所需的用水量
lead-free	無鉛	由美國環保局 (EPA) 在《安全飲用水法》（Safe Drinking Water Act）中規定的標籤，允許在焊料、焊劑、管道、管件和井泵中出現少量鉛
least-risk pesticide	最低危害殺蟲劑	一種使用舊金山危險等級制度（San Francisco Hazard Ranking System）登記過的三級（最低毒性）殺蟲劑，或者符合「三藩市殺蟲劑危險篩查協議」（San Francisco Pesticide Hazard Screening Protocol）的要求，且作爲在建築住戶無法接近的區域中使用的獨立誘餌或者裂縫和裂隙處理劑出售的殺蟲劑。滅鼠藥從未被視爲最低風險殺蟲劑。

Term (English)	繁體中文名詞	中文名詞定義
length of stay	住院時長	個人作為住院患者在衛生保健設施中逗留的時間
life-cycle assessment	生命週期評估	對產品整個生命週期中環境影響的評估，符合 ISO 14040-2006 和 ISO 14044-2006 的定義
life-cycle inventory	生命週期清單	一個定義材料或構件生命週期中每個步驟環境影響（輸入和輸出）的資料庫。該資料庫視具體國家和地區而異。
light pollution	光污染	是指來自建築基地、照向天空或投射到基地外的會產生眩光的廢光。廢光並不會增加夜間安全性或利於保安工作，不具有實用性，並且還會不必要地消耗能源。
light rail	輕軌	使用 2 節或 3 節車廂的列車在道路用地中提供的公共交通服務，通常與其他交通模式分隔開。站間距離傾向於 1/2 英里或更遠，最大行駛速度一般為 40-55 英里／小時（65-90 公里／小時）。輕軌線路一般為 10 英里（16 公里）或更長。
light trespass	光侵擾	因為數量、方向或光譜屬性而顯得刺眼且不必要的照明。光侵擾可造成煩擾、不適、注意力分散，或使視力受損。
load shedding	切負荷	公用事業公司為降低系統負載的特意執行措施。切負荷通常在緊急時期進行，如在容量不足、系統不穩定或電壓控制時。
long-term bicycle storage	長效自行車存放	便於居民和員工使用的自行車停放處，有頂篷以防止自行車受到雨雪影響

Term (English)	繁體中文名詞	中文名詞定義
low-cost im-provement	低成本改進	一種運營改進，如維修、升級、員工培訓或挽留。在 LEED 中，專案團隊根據設施資源和運營預算來確定低成本改進的合理上限。
low-impact development (LID)	低衝擊開發（LID）	一種管理雨水徑流的方法，強調基地內自然特點，以通過恢復流域內的自然土地植被水情和處理靠近源頭的徑流來保護水質。示例包括更好的基地設計原則（如儘量減少土地破壞、保留植被、儘量減少非滲透表面），以及設計實踐（雨水花園、植被窪地和緩衝區、透水路面、雨水收集和土壤改良）。這些都是工程實踐，可能需要專業設計輔助。
makeup water	補充水	送入冷卻塔系統或蒸發式冷凝器的水，用於替換通過蒸發、漂流、溢流或其他原因流失的水
manage (rain-water) on site	基地內雨水管理	收集和保留指定量的雨水，以模擬自然水文功能。雨水管理的範例包括涉及蒸散、滲透、捕獲和再利用的策略。
master plan boundary	總平面圖邊界	基地總平面圖的界限。總平面圖邊界包括專案區域，還可包括 LEED 專案邊界以外的所有相關建築和基地。總平面圖邊界考慮了未來基地的可持續使用、擴張和收縮。
mean lumen output a mea-surement	平均流明輸出	源自行業標準的一種對光源發出的光進行測量的指標，與鎮流器因數為 1.0 的暫態啟動鎮流器一起在 40% 的燈具使用壽命下測得（T-5 燈除外，它使用程式啟動鎮流器）。
medical fur-nishing	醫療陳設	設計用於醫療保健的傢俱。其中包括手術臺；手術推車、耗材推車和移動設備推車；升降和轉移輔助裝置；耗材櫃推車和架子，以及跨床桌。

Term (English)	繁體中文名詞	中文名詞定義
metering control	定量調節器	限制水流時間的調節器，一般是一種手動打開和自動關閉的設備，最常安裝在衛生間的水龍頭和淋浴器上
mixed paper	混合紙	白紙和彩色紙、信封、表格、資料夾、寫字板、傳單、麥片盒子、包裝紙、目錄、雜誌、電話簿和照片
modular and movable casework	模組化（可移動）台櫃	被設計爲可以輕鬆安裝、移動和重新配置的貨架和櫥櫃。在零售環境中，可以移動但使用機械緊固系統半永久性固定以便於日常經營的傢俱被視爲傢俱，不是基本建築元素（如，通過螺栓固定到地板上的桌子和展架，或者固定到牆壁的貨架）。
mounting height	安裝高度	地平面（或工作面）與光源（燈具）底部之間的距離；光源安裝的高度。（引自 Light a Home）
movable furniture and partitions	可移動傢俱和隔斷	用户無需工具或特殊行業和設施管理人員的協助即可移動的傢俱
multitenant complex	多租户綜合體	爲開發商店、餐館和其他業務而總體規劃的基地。零售商可以共用某些服務和公共區域。
NAED code	NAED 代碼	一個唯一的 5 位或 6 位數，用於確定具體的燈具，由全國電氣分銷商協會 (National Association of Electrical Distributors) 使用
native vegetation	本地植物	沒有人類直接或間接活動的特定區域、生態系統和棲息地中出現的本地物種。本地物種隨著該地區的地理、水文和氣候而進化。它們還會出現在社區中；即，他們與其他物種一同進化。因此，這些社區可爲多種其他本地野生動物提供棲息地。北美本地的物種一般被認爲是早于歐洲移民出現在北美大陸上。也被稱爲本地植栽。

Term (English)	繁體中文名詞	中文名詞定義
natural refrigerant	天然製冷劑	一種非人造的製冷化合物。相比人造化學製冷劑，這種物質不大可能對大氣造成破壞。包括水、二氧化碳和氨。
Natural Resources Conservation Service (NRCS) soils delineation	自然資源保護局（NRCS）土壤描繪	一項基於美國的土壤調查，顯示基地內不同土壤類型的邊界和特殊土壤特性
natural site hydrology	自然基地水文	水的出現、分布、移動和平衡的自然土地覆被作用
net usable program area	淨可用計畫區域	專案中可用於容納專案計畫的所有內部區域之和。不包括用於建築設備、垂直迴圈或結構元件的區域。
non-inpatient area	非住院病人區域	公共空間、診斷或治療區域、門診部，或衛生保健設施中不用於已接受住院護理的個人的其他任何空間
nonpotable water	非飲用水	不符合飲用水標準的水
non-regularly occupied space	非常規使用空間	人們經過或用於重點活動的區域，持續時間平均每人每天不到一小時。一小時的時間範圍是連續的，且應基於典型住戶使用該空間的時間。對於不是每天使用的空間，這一小時的時間範圍應基於典型住戶在該空間開放使用時在其內消耗的時間。
nonwater toilet systems	無水馬桶系統	容納人類排泄物並通過微生物方式進行處理的乾燥衛生器具和裝置
nonwater urinal	無水小便器	不用水沖洗的衛生器具，防臭瓣中有一層浮液浮在尿液上面，阻隔下水道氣體和異味

Term (English)	繁體中文名詞	中文名詞定義
occupant control	住戶控制裝置	空間中的人可以直接接觸和使用的系統或開關。包括工作照明、開啓式開關和百葉窗。溫度感測器、光電感測器或中央控制系統無法被住戶控制。
occupiable space	可使用空間	用於人類活動的封閉空間，不包括主要用於其他用途的空間，如儲藏室和設備間，以及僅偶爾占用或短期佔用的空間 (ASHRAE 62.1-2010)
occupied space	使用空間	用於人類活動的封閉空間，不包括主要用於其他用途的空間，如儲藏室和設備間，以及僅偶爾佔用或短期佔用的空間。使用空間根據使用時間可進一步劃分爲經常使用或不經常使用的空間，根據住戶數量可分爲單人空間和多住戶空間，根據空間中的住戶密度分爲密集使用和非密集使用空間。
ongoing consumable	持續消耗品	單位成本較低且在業務經營期間定期使用和補充的產品。包括紙張、碳粉、粘合劑、電池和辦公用品。也被稱爲持續採購品。
on-site wastewater treatment	基地內汙水處理	運輸、存放、處理和處置專案基地產生的廢水
open-grid pavement system	開放鏈式鋪裝系統	鬆散材料組成的路面，由結構更加結實的網格或網帶提供支撐。透水混凝土和多孔瀝青不被認爲是開放式網格，而是有邊界材料。無邊界的鬆散材料不像有邊界的緊實材料那樣傳遞和儲存熱量。
operations and maintenance (O&M) plan	運營和維護計畫	指定主要系統運營參數和限制、維護程式和安排，以及爲說明正確操作和維護所獲准的排放控制設備或系統所需的記錄方法的計畫

Term (English)	繁體中文名詞	中文名詞定義
ornamental luminaire	景觀照明	用於迴圈網路照明部分的光源，除了提供有效的街道照明，也兼具觀賞功能，並且具有裝飾性或古典外觀
outpatient	門診病人	在 24 小時或更長時間內沒有住院，但是期間去過醫院、診所或相關醫療機構進行診斷或治療的患者
owner's project requirements (OPR)	業主專案要求（OPR）	由業主確定的，詳述了對專案成功十分重要的想法、理念和條件的書面文件
patient position	病人位置	病床、輸液椅、康復室開間，或病人接受臨床護理的其他位置
peak demand	峰值需求	特定時間點或某個時期的最大電力負荷
peak watering month	峰值用水月份	蒸散和降雨量之間出現最大虧缺的月份。基地區域內的植物最需要補充水分的月份，一般是盛夏中的月份。(可持續基地措施)
permanent interior obstruction	永久室內阻礙	在沒有工具或特殊行業和設施管理人員的幫助下，使用者無法移動的結構。包括實驗室罩、固定隔斷、可拆卸的不透明全高或半高隔斷、一些展架和設備。
permanent peak load shifting	永久性峰值負載轉移	將能耗轉移到能源需求較低且因此能源費用較便宜的非高峰期
permeable pavement	可滲透鋪裝	可以讓雨水徑流滲入地下的鋪設表面
persistent bio-accumulative toxic chemical	永久性生物累積毒素（PBT）	對人類和環境帶來長期風險的物質，因為會長期存留在環境中，並且濃度會隨著食物鏈層級的增加而增加，可以轉移到距離污染源很遠的地區。往往存在時間越長，這些物質的濃度就越高，對生態系統的破壞也越大。有

Term (English)	繁體中文名詞	中文名詞定義
		關永久性生物累積毒素的資訊，請參見美國環保局 (EPA) 網站 www.epa.gov/pbt/。
place of respite	療養區	可以使醫療保健病人、訪客和員工受到自然環境健康益處的區域。（引自 Green Guide for Health Care Places of Respite Technical Brief）
plug load or receptacle load	插座負荷	所有通過牆壁插座連接到電氣系統的設備消耗的電流。
postconsumer recycled content	消費後回收物質含量	家庭或商業、工業和機構設施作為產品最終使用者產生的，不可再用於其原有用途的廢棄物
potable water	飲用水	符合或超出美國環保局飲用水水質標準（美國以外為當地同等標準），且經州或地方管轄當局批准用於人類消費的水；可通過水井或市政供水系統供給
power distribution unit output	配電裝置輸出	來自為資訊技術（IT）設備分配電能以及為其服務的設備的電能。配電裝置（PDU）輸出不包括 PDU 中任何轉換的效率損失，但可以包括安裝在 IT 機架中的下游非 IT 輔助設備，如風扇。如果 PDU 系統支援非 IT 設備（如，電腦房空調裝置、電腦房空氣處理裝置、行級製冷器），必須計量該設備的用電量並從 PDU 輸出讀數中減去。計量方法應與用電效率（PUE）類別要求的計量方法一致（如，PUE 類別 1、2 和 3 的連續消耗計量）。
power utilization effectiveness (PUE)	電力使用有效性（PUE）	資料中心電源使用效率指標；尤其是，由計算設備而不是冷卻或其他開銷使用的電量
powered floor maintenance equipment	電動地板維護設備	電動和電池供電的地板減振墊和磨光器。不包括濕應用中使用的設備

Term (English)	繁體中文名詞	中文名詞定義
preconsumer recycled content	消費前回收物質含量	在製造過程中由廢物流轉化而成的物質，由材料（重量）百分比確定。例如：刨花、鋸屑、甘蔗渣、核桃殼、下腳料、修剪材料、過量發行的出版物及陳舊存貨，不包括可通過與生產材料相同的過程予以回收的返工、粉碎再生材料或廢料（ISO 14021）。之前被稱爲工業用後含量。
preferred parking	優先車位	最靠近建築主入口的停車位（指定的殘疾人車位除外）。對於員工停車，是指最靠近員工使用的入口的停車位。
premature obsolescence	過早廢棄	棄用或停用使用壽命超出設計壽命的構件或材料。例如，潛在壽命爲30年的材料被特意設計爲只使用15年，這樣其剩餘15年的使用壽命就有可能會被浪費掉。相反，使用壽命與預期使用相同的構件可以發揮出最大使用潛力。
previously developed	先前已開發	通過鋪裝、建設和／或土地使用而經過改變，一般都有所需的監管啓動許可（現在或過去存在改變）。先前未開發的土地，以及被當前或歷史上的清理或塡埋、農業或林業用途改變的景觀，或者保留的天然區域均被視爲未開發的土地。之前開發許可發放的日期構成了先前開發的日期，但是許可發放本身並不構成先前開發。
previously developed site	先前已開發基地	基地在專案之前包括至少75%的先前已開發土地
prime farmland	基本農田	由美國農業部的自然資源保護局確定（一種基於美國的方法，爲高產土壤制定標準）的具有生產食物、飼料、草料、纖維和油料作物所需最佳物理和化學特性組合，並可用於這些用途的土地。有關基本農田標準的完整說明，請參見美國聯邦法規第7部，第6卷，400到699部分，657.5小節。

Term (English)	繁體中文名詞	中文名詞定義
private meter	私有水錶	一種測量水流的設備，安裝在公用水錶的下游，或者作爲由建築管理團隊維護的基地內供水系統的一部分
process energy	作業用能	爲了支援生產、工業或商業流程而消耗的電能，而不是爲建築住戶提供空調空間及保持舒適所消耗的電能。可包括製冷、烹飪和食物準備、衣物洗滌設備以及其他主要輔助電器。（ASHRAE）
process load or unregulated load	作業負荷或非常規負荷	由於作業用能消耗或釋放而產生的建築負荷（ASHRAE）
process water	作業用水	工業流程及建築系統（如冷卻塔、鍋爐和冷卻器）用水。此外，也可以指操作流程中使用的水，如洗碗、洗衣服，以及制冰。
product (permanently installed building product)	產品（永久安裝的建築產品）	作爲準備好安裝的成品元素或者在基地內組裝的其他物品的構件到達專案基地的物品。產品單位由專案中使用的功能要求定義；這包括物理構件和爲永久安裝的建築產品的預期功能服務所需的服務。此外，對於規範中的類似產品，每種都可作爲一個單獨的產品。
public water supply (PWS)	公共供水（PWS）	通過管道或建造的其他運輸工具爲公眾供水以使用於人類消費的系統。若要被視爲公共系統，這種系統必須有至少 15 個服務連接或者爲至少 25 個人提供定期服務。（引自美國環保局（EPA））
rainwater harvesting	雨水收集	雨水的收集、轉移和存儲，以便將來用作有利用途。一般來說，雨桶或蓄水池用於儲水；其他構件包括集水區表面及運輸系統。收集的雨水可用於灌溉。

Term (English)	繁體中文名詞	中文名詞定義
raw material	原材料	製作產品的基本物質，如混凝土、玻璃、石膏、磚石、金屬、再生材料（如塑膠和金屬）、油（石油聚乳酸）、石材、農業纖維、竹子和木材
reclaimed water	再生水	經過處理和淨化以便重複利用的廢水
recycled content	回收物質含量	根據國際標準組織文件，ISO 14021—環境標誌和聲明—自我環境聲明（II類環境標誌）進行定義
reference evapotranspiration rate	參考蒸散率	沒有水分限制的特定植被表面失去的水量。高 120 mm 的草皮是參考植被。
reference soil	參考土壤	按照自然資源保護局土壤調查（美國以外地區為當地的同等調查）的描述，為專案基地本地的土壤，或者專案區域內未受干擾、有原生植被、地形且土壤紋理與專案基地相似的原生土壤。對於無既有土壤的專案基地，參考土壤被定義為專案區域內未受干擾，且為與新專案預期物種類似的適當原生植物物種提供支援的原生土壤。
refurbished material	翻新材料	已完成其生命週期並準備在不大幅更改其形態的情況下再利用的物品。翻新包括整修、修復、恢復，或通常為改善外觀、性能、品質、功能或產品價值。
regularly occupied space	常用空間	一個或多個人在建築中工作、學習或進行其他重點活動時經常（平均每人每天超過一小時）坐立的區域。一小時的時間範圍是連續的，且應基於典型住戶使用該空間的時間。對於不是每天使用的空間，這一小時的時間範圍應基於典型住戶在該空間開放使用時在其內消耗的時間。

Term (English)	繁體中文名詞	中文名詞定義
regularly used exterior entrance	常用外部入口	常用是指進入建築。包括主建築入口以及與停車場、地下停車場、地下通路或外部空間相連的任何建築入口通道。不包括非典型入口、緊急出口、中庭、廣場之間的連接以及內部空間。
regulated load	調節負荷	在 ANSI/ASHRAE/IES 標準 90.1-2010 中有強制性或指令性要求的任何建築最終用途
remanufactured product	再製造產品	經過維修或調整並退回進行檢修的物品。再製造產品預計可以和新產品一樣使用。
renewable energy	可再生能源	不因使用而耗盡的能源。例如：太陽能、風能、小型（低影響）水力發電，以及地熱能與潮波系統。
renewable energy credit (REC)	可再生能源認證（REC）	證明單位電量是源自可再生資源的一種可交易商品。REC 與電力本身分開而單獨出售，從而使傳統電力的用戶得以購買綠色電力。
reuse	再利用	按照與原始應用相同或相關的功能重新使用材料，從而延長了材料的使用壽命，避免被丟棄。再利用包括回收和重新使用從既有建築或施工基地回收的材料。也稱為重新利用。
reused area	再利用面積	專案原始狀況中既有且仍會包括在最終設計之內的建築結構、核心以及外圍護結構的總面積
revenue-grade meter	收費等級表	設計用於滿足法規或法律要求的嚴格精確度標準的測量工具。公用事業計量表通常被稱為收費等級表，因為其測量直接導致對客戶收費。

Term (English)	繁體中文名詞	中文名詞定義
rideshare	共乘	一種公共交通服務,是指多人乘坐一輛至少能坐 4 個人的乘用車或小貨車出行。可包括人力運輸工具,必須容納至少 2 個人。必須包括封閉的乘客座位區、固定路線服務、固定費用結構,保持日常運營,以及能夠搭乘多人。
salvaged material	廢舊利用材料	從既有建築或施工現場回收並重複使用的建築構件。常見的可重新利用的建築構件包括結構樑柱、地板、門、櫥櫃、磚及裝飾構件。
school authority	學校當局	負責有關學校運營、地區、人員、財務和未來發展等決策制定的機構。包括學校董事會、當地政府和宗教機構。
Scope 1 emissions	第 1 範圍排放	從實體擁有或控制的來源直接排放溫室氣體,如基地內化石燃料燃燒產生的排放
Scope 2 emissions	第 2 範圍排放	與從公用事業供應商購買以供實體使用的電力、加熱 / 製冷或基地外蒸氣的產生相關的間接溫室氣體排放
sDA300/50%	sDA300/50%	在至少 50% 的分析期中,整個分析區內滿足或超過 300 勒克斯的分析點的百分比
server input	伺服器輸入	在將 IT 設備連接到電力系統時(如,電源插座)測量的資訊技術(IT)負載。伺服器輸入收集 IT 設備的實際功率負載,不包括配電損失和非 IT 負載(如,安裝在機架上的風扇)。
service life	使用壽命	為了生命週期評估,假設建築、產品或組合材料可使用的時間長度
shared multioccupant space	共用多住戶空間	集會地點,或者住戶追求重疊或協作任務的地點

Term (English)	繁體中文名詞	中文名詞定義
shell space	殼體空間	設計用於在將來進行擴建的區域。殼體空間被建築外圍護結構封閉起來，但是無塗層。
short-term bicycle storage	短時自行車存放	非封閉式自行車停車，一般由訪客使用，持續 2 個小時或更短時間。
simple box energy modeling analysis	簡單箱體能源模擬分析	（也被稱爲「建築量體模型的能源分析」）一種簡單的基本情況能源分析，告知團隊建築可能的能耗分配，用於評估潛在的專案能源策略。簡單箱體分析使用基本、示意性的建築形式。
site assessment	基地評估	評估一個基地地上和地下的特性，包括其結構、地質和水文特性。基地評估通常有助於確定是否已發生污染，以及任何污染物排放的程度和濃度。補救決定依靠基地評估期間生成的資訊。
site master plan	基地總平面圖	專案以及相關（或潛在相關）建築和基地的整體設計或開發理念。總平面圖考慮了未來基地的可持續使用、擴張和收縮。基地總平面圖通常以建築規劃（如適用）、計畫的各階段的基地圖紙和敘述性描述進行說明。
soft space	軟空間	功能可以輕易改變的區域。例如，醫院行政辦公室可以移動，從而這個軟空間可以變爲實驗室。相反，具有專業設備和基礎設施的實驗室很難改變位置。
softscape	軟景觀	包括活的園藝元素的景觀元素
solar garden/ community renewable energy system	太陽能花園／社區可再生能源系統	使用虛擬淨測量值的共用太陽能電池陣列或其他可再生能源系統，擁有連接了電網的用戶，這些用戶因使用可再生能源而獲得得分點（引自 solargardens.org）

Term (English)	繁體中文名詞	中文名詞定義
solar reflectance (SR)	太陽能反射（SR）	被表面反射的太陽能所占的比例，範圍爲 0 到 1。黑色油漆的日光反射係數爲 0；白色油漆（二氧化鈦）的日光反射係數爲 1。其測量使用的標準技術是分光光度測量法，通過積分球來確定每個波長的反射。然後按照 ASTM 標準 E903 和 E892 的要求，使用標準太陽光譜，通過平均處理來獲得平均反射率。
solar reflectance index (SRI)	太陽能反射指數（SRI）	用於測量建築表面通過反射太陽輻射並發出熱輻射，在陽光下保持涼爽的能力。定義標準爲，標準黑色表面（初始太陽能反射率爲 0.05，初始熱反射率爲 0.90）的初始 SRI 爲 0，標準白色表面（初始太陽能反射率爲 0.80，初始熱反射率爲 0.90）的初始 SRI 爲 100。若要計算指定材料的 SRI，需通過涼屋面評級委員會標準（CRRC-1）獲取其太陽能反射率和熱反射率。SRI 根據 ASTM E 1980 計算。老化 SRI 的計算基於太陽能反射率和熱反射率的老化測試值。
sound-level coverage	音量覆蓋範圍	一組統一標準，用於確保對空間中所有住戶一致的可聞頻率清晰度和方向
source reduction	減少資源消耗	減少帶進建築的不必要的材料數量，以產生更少的廢棄物。例如，購買包裝更少的產品就是一種減少資源消耗的策略。
spatial daylight autonomy (sDA)	空間日照自足指數（sDA）	一種描述室內環境中環境日光水準年充分性的指標。被定義爲在每年指定的一段運營時間內符合最低日光照度水準的分析區域（執行計算的區域，一般橫跨整個空間）的百分比（即，符合 Reinhart & Walkenhorst, 2001 的日照自足值）。照度水準和時間部分作爲下標包含在內，與 sDA300,50% 中一樣。sDA 值被表示爲面積百分比。

Term (English)	繁體中文名詞	中文名詞定義
speech privacy	話語私密性	話語讓被動聽者難以聽懂的狀態（ANSI T1.523-2001）
speech spectra	語譜	對於人類語言來說爲作爲頻率函數的聲能分佈
streetcar	有軌電車	一種使用小型、獨立軌道車輛的公共交通服務。站間距離一致且很短，範圍從每個街區到 1/4 英里，運行速度一般爲 10-30 英里／小時 (15-50 公里／小時)。有軌電車路線一般長達 2-5 英里（3-8 公里）。
structure	建築結構	承載垂直或水準負載，且被認爲架構穩定、無危險的構件（如牆、屋面和地板）
systems manual	系統手冊	提供瞭解、運營和維護建築中的系統及組合材料所需的資訊。擴展了傳統運營與維護檔的範圍，包括多個在調試過程中制定的檔，如業主專案要求、運營與維護手冊和運營順序。
Technical Release (TR) 55	Technical Release (TR) 55	由前美國農業部水土保持局（USDA Soil Conservation Service）開發的一種水文測定方法，通過對流域建模以計算暴雨徑流量、峰值排水率、水點陣圖和儲存量
thermal emittance	熱發射率	相同溫度下，樣品發出的輻射熱通量與黑體輻射源發出的輻射熱通量的比值（引自涼屋面評級委員會）
three-year aged SR or SRI value	三年期 SR 或 SRI 値	在露天條件下三年之後測得的太陽能反射率或太陽能反射指數
time-of-use pricing	按時段計價	一種讓客戶在高峰時期支付較高的價格來使用公用設施，反之在非高峰時期支付較低價格的規定

Term (English)	繁體中文名詞	中文名詞定義
undercover parking	非露天停車場	位於地下、露天平臺下、屋面下或建築下方的停車場
uninterruptible power supply (UPS) output	不斷電供應系統（UPS）輸出	由一種可以在停電時保持資訊技術（IT）設備正常工作的裝置提供的電能。UPS 輸出不包括來自裝置本身的效率損耗，但包括下游配電元件的損耗，這些元件包括配電裝置，還包括安裝在 IT 機架中的非 IT 輔助設備，如風扇。如果 UPS 系統支援非 IT 設備（如，電腦房空調裝置、電腦房空氣處理裝置、行級製冷器），必須計量該用電量並從 UPS 輸出讀數中減去。計量方法應與用電效率（PUE）類別要求的計量方法一致（如，PUE 類別 1、2 和 3 的連續消耗計量）。
universal waste	常見廢棄物	很容易就能買到且常用的危險物品。包括電池、殺蟲劑、含汞設備和燈泡。參見 epa.gov/osw/hazard/wastetypes/universal/index.htm。
unoccupied space	非人員使用空間	一種設計用於設備、機械或儲存而不是人類活動的區域。設備區域只有在偶爾取放設備的情況下才會被認為是未使用的。
upstream equipment	上游設備	與區域能源系統（DES）相關的加熱或冷卻系統或控制裝置，但不是與 DES 之間的熱連接或介面。上游設備包括熱能轉化廠，以及在將熱能運送到專案建築或場址期間所有的傳輸和分配設備。
USDA Organic	USDA 有機認證	美國農業部為所含成分中至少 95%（不包括水和鹽）是使用非合成化學物質、抗生素或激素生產的產品授予的認證。其他成分必須由經過 USDA 批准的非農業物質或不以有機形式市售的農業產品組成。

Term (English)	繁體中文名詞	中文名詞定義
vertical illuminance	垂直照度	在垂直表面上的一點，或出現在垂直平面上的一點計算的照度水準。
vision glazing	視野眩光	外部窗戶中可以看到外部或內部的玻璃。觀景窗必須能夠清晰地看到外面，並且不得被扭曲色彩平衡的玻璃料、纖維、有圖案的玻璃或添加的色彩遮擋。
walking distance	步行距離	行人必須在安全舒適的環境中，在出發點和目的地之間無障礙地在連續的人行道網路上、露天人行道、人行橫道或類似的行人設施上行走的距離。步行距離必須從所有建築用戶都可進入的入口開始算起。
waste diversion	廢棄物轉化	通過焚化爐或填埋場以外的方式處理廢棄物的一種管理活動。包括再利用和回收。
waste-to-energy	從廢物到能量	將不可回收的廢料通過多個流程轉換爲可用的熱能、電能或燃料，這些流程包括燃燒、氣化、熱裂解、厭氧消化、垃圾填埋氣（LFG）回收
water body	水體	溪流（一級或更高，包括間歇性河流）小溪、河、運河、湖、出海口、海灣或者海洋的地表水。不包括灌溉水渠。
water budget	用水預算	用於計算建築和相關地面所需水量的項目特定方法。預算考慮室內、室外、作業和補充水需求，以及任何基地內供應，包括預計的降水。用水預算必須與具體的時段（如周、月或年）和水量（如 kGal 或升）相關。
wet meter	水錶	一種安裝在水管內部以記錄流過的水量的設備

Term (English)	繁體中文名詞	中文名詞定義
wetland	濕地	足以維持以一定的頻率和持續時間被地表水或地下水淹沒或滲透，以及在正常環境下能夠維持眾多適應飽和土壤條件的植物生存的區域。濕地一般包括沼澤、濕地、泥沼以及類似的區域，但不包括灌溉水渠（除非劃定為相鄰濕地的一部分）。
wood	木材	基於植物且符合森林管理委員會認證條件的材料。包括竹子和棕櫚（單子葉植物）以及硬木（被子植物）和軟木（裸子植物）
xeriscaping	旱生園藝	無需常規灌溉的景觀
yard tractor	堆置場牽引車	主要用於方便卡車拖車和其他類型大型集裝箱從基地的一個區域移動到另一個區域的車輛。不包括叉車。也被稱為牽引車、堆場拖車、多用途拖拉機裝置、牽引頭、堆場搬運車或堆場拖車。
zero lot line project	零紅線專案	建築占地通常或幾乎與場址邊界一致的地點

12.3 繁體中文詞彙（Chinese Technical Glossary）

Acid rain: the precipitation of dilute solutions of strong mineral acids, formed by the mixing in the atmosphere of various industrial pollutants (primarily sulfur dioxide and nitrogen oxides) with naturally occurring oxygen and water vapor.	酸雨：成分為強礦物酸稀釋液的降雨。所含礦物酸由大氣中各種工業污染物（主要為二氧化硫和氮氧化物）與自然存在的氧氣及水蒸氣混合形成。

Adapted plants: nonnative, introduced plants that reliably grow well in a given habitat with minimal winter protection, pest control, fertilization, or irrigation once their root systems are established. Adapted plants are considered low maintenance and not invasive.	可適應性植物：此類非本地植物在根莖系統建立後，對於冬季防護、害蟲防治、施肥或灌溉的要求較低，且在給定的生活環境下生長良好。可適應性植物被認為是維護需求低且不具入侵性的植物。
Adaptive reuse: designing and building a structure in a way that makes it suitable for a future use different than its original use. This avoids the use of environmental impact of using new materials.	適應性再利用：設計和建造一種使建築物改變其原有用途，以適應其未來之使用功能，避免由於使用新材料對環境造成影響。
Air quality standard: the level of a pollutant prescribed by regulations that are not to be exceeded during a given time in a defined area (EPA).	空氣品質標準：法規規定的對於指定區域在給定時間範圍內不可超出的污染物水準（美國環保局 EPA）。
Albedo: the reflectivity of a surface, measured from 0 (black) to 1 (white).	Albedo：物體表面對光線的反射率，範圍介於 0（黑色）至 1（白色）之間。
Ambient temperature: the temperature of the surrounding air or other medium (EPA).	環境溫度：周圍空氣或其它介質的溫度（美國環保局 EPA）。
Alternative fuel vehicle: a vehicle that uses low-polluting, nongasoline fuels, such as electricity, hydrogen, propane or compressed natural gas, liquid natural gas, methanol, and ethanol. In LEED, efficient gas- electric hybrid vehicles are included in this group.	替代燃料汽車：使用低污染非汽油燃料（如電力、氫氣、丙烷或壓縮天然氣、液化天然氣、甲醇及乙醇）的汽車。在 LEED 體系中，此類別還包括高效氣電混合動力汽車。
ASHRAE: American Society of Heating, Refrigerating and Air- Conditioning Engineers.	ASHRAE：美國採暖、製冷與空調工程師學會。

Bake-out: a process used to remove volatile organic compounds (VOCs) from a building by elevating the temperature in the fully furnished and ventilated building prior to human occupancy.	烘乾：入住以前，在配套齊全且通風良好的建築的室內通過提高溫度來去除建築內揮發性有機化合物（VOC）的過程。
Baseline versus design or actual use: the amount of water that the design case or actual usage (for existing building projects) conserves over the baseline case. All Water Efficiency credits use a baseline case against which the facility's design case or actual use is compared. The baseline case represents the Energy Policy Act of 1992 (EPAct 1992) flow and flush rates and the design case is the water anticipated to be used in the facility.	基準值與設計或實際使用的對比：與基準值案例相比，設計案例或實際使用（針對現有建築專案）所節省的水量。所有節水得分點均將基準值案例與設施的設計案例或實際用途進行比較。基準值案例代表了 1992 年能源政策法案（EPAct1992）規定的流速和沖洗速度，設計案例為設施的預計用水量。
Biodegradable: capable of decomposing under natural conditions (EPA).	可生物降解：在自然條件下可分解（美國環保局 EPA）。
Biodiversity: the variety of life in all forms, levels, and combinations, including ecosystem diversity, species diversity, and genetic diversity.	生物多樣性：生物在形式、層次及組合上的多樣性，包括生態系統多樣性、物種多樣性及遺傳多樣性。
Biomass: plant material from trees, grasses, or crops that can be converted to heat energy to produce electricity.	生物質：可轉換為熱能以用於產生電能的樹木、草類或作物植物體。
Bioswale: a stormwater control feature that uses a combination of an engineered basin, soils, and vegetation to slow and detain stormwater, increase groundwater recharge, and reduce peak stormwater runoff.	生態草溝：一種結合了人工建造的盆地、土壤及植被的雨水控制設施，用以減緩和保留雨水，增加地下水補給並減少高峰雨水徑流。

Blackwater: wastewater from toilets and urinals; definitions vary, and wastewater from kitchen sinks (perhaps differentiated by the use of a garbage disposal), showers, or bathtubs is considered blackwater under some state or local codes.	污水：馬桶與小便池廢水；定義視情況有所差異。根據某些州或地方法規，廚房水槽（或可根據垃圾處理進行區分）、淋浴或浴缸廢水被視為黑水。
British thermal unit (Btu): the amount of heat required to raise the temperature of one pound of liquid water from 60° to 61° Fahrenheit. This standard measure of energy is used to describe the energy content of fuels and compare energy use.	英制熱量單位（**Btu**）：將一磅液態水由 60°F 加熱至 61°F 所需的熱量。這種能量度量標準用於說明燃料所含的能量以及比較能量使用。
Brownfield: previously used or developed land that may be contaminated with hazardous waste or pollution. Once any environmental damage has been remediated, the land can be reused. Redevelopment on brownfields provides an important opportunity to restore degraded urban land while promoting infill and reducing sprawl.	褐地：經使用或開發過的、可能被危險廢棄物或污染物污染的土地。一旦任何環境損害得以修復，該土地即可重新投入使用。褐地的再開發提供了恢復退化城市土地的重要機遇，同時可促進填充式開發並減少無計畫擴張。
Building commissioning: a systematic investigation comparing building performance with performance goals, design specifications, and the owner's requirements.	建築調試（功能驗證）：將建築性能與性能目標、設計細節及業主要求進行比較的系統性調查。
Building density: the floor area of the building divided by the total area of the site (square feet per acre).	建築容積率：總建築面積除以基地總面積（平方英尺／英畝）。
Building envelope: the exterior surface of a building—the walls, windows, roof, and floor; also referred to as the building shell.	建築皮（外殼）結構：建築外表面——外牆、窗戶、屋頂與樓面；也稱之為建築外殼。

Building footprint: the area on a project site that is used by the building structure, defined by the perimeter of the building plan. Parking lots, landscapes, and other nonbuilding facilities are not included in the building footprint.	建築占地：建築結構在專案基地上占用的面積，根據建築平面圖的周長確定。建築占地不包括停車場、景觀和其它非建築設施。
Built environment: a man-made environment that provides a structure for human activity.	建築環境：為人類活動提供建築構築物的人造環境。
Byproduct: material, other than the principal product, generated as a consequence of an industrial process or as a breakdown product in a living system (EPA).	副產品：在工業過程中生成的，或由生命系統分解產物而產生的除主產品以外的材料（美國環保局）。
Carbon dioxide concentration: an indicator of ventilation effectiveness inside buildings; CO_2 concentrations greater than 530 parts per million (ppm) above outdoor conditions generally indicate inadequate ventilation. Absolute concentrations of greater than 800 to 1,000 ppm generally indicate poor air quality for breathing. CO_2 builds up in a space when there is not enough ventilation.	二氧化碳濃度：建築內通風效率的指標；一般情況下，若室內 CO_2 濃度高於戶外條件下的百萬分之 530（ppm），則被視為通風不足。若絕對濃度高於 $800\sim1000$ppm，則一般情況下被視為呼吸空氣品質較差。通風不足時，CO_2 會在空間內累積。
Carbon footprint: a measure of greenhouse gas emissions associated with an activity. A comprehensive carbon footprint includes building construction, operation, energy use, building-related transportation, and the embodied energy of water, solid waste, and construction materials.	碳足跡：對與某一活動相關的溫室氣體排放量的度量。全面的碳足跡包括建築施工、運營、能源消耗、與建築相關的交通，以及水、固態廢棄物與建築材料的蘊藏能量。
Carbon neutrality: emitting no more carbon emissions than can either be sequester or offset.	碳中和：除可隔離或可抵消的碳排放以外，不再有其它碳排放。

Charrette: intense workshops designed to produce a specific deliverables.	專家研討會：旨在取得特定交付成果的集中研討會。
Chiller: a device that removes heat from a liquid, typically as part of a refrigeration system used to cool and dehumidify buildings.	冷水機組：去除液體熱量的設備。通常為建築製冷及除濕用製冷系統的一部分。
Chlorofluorocarbon (CFC): an organic chemical compound known to have ozone-depleting potential.	氯氟化碳（**CFC**）：一種已知的具備臭氧消耗潛能的有機化合物。
Closed system: a system that exchanges minimal materials and elements with its surroundings; systems are linked with one another to make the best use of byproducts.	封閉系統：一種與周圍環境交換最少量材料及元素的系統；各系統之間彼此相連，以充分利用副產品。
Commissioning (Cx): the process of verifying and documenting that a building and all of its systems and assemblies are planned, designed, installed, tested, operated, and maintained to meet the owner's project requirements.	功能驗證（**Cx**）：也稱為調試，檢驗與記錄一棟建築及其所有系統與元件如何在規劃、設計、安裝、測試、操作及維護上滿足業主專案要求的過程。
Commissioning plan: a document that outlines the organization, schedule, allocation of resources, and documentation requirements of the commissioning process.	功能驗證計畫：簡要說明調試過程組織、進度計畫、資源配置及檔要求的檔。
Commissioning report: a document that details the commissioning process, including a commissioning program overview, identification of the commissioning team, and description of the commissioning process activities.	功能驗證報告：詳細說明調試過程，包括調試計畫概述、調試團隊識別及調試過程活動描述的檔。

community connectivity: the amount of connection between a site and the surrounding community, measured by proximity of the site to homes, schools, parks, stores, restaurants, medical facilities, and other services and amenities. Connectivity benefits include more satisfied site users and a reduction in travel associated with reaching services.	社區關聯性：建築基地與周邊社區之間的關聯程度。通過基地與住宅、學校、公園、商店、餐廳、醫療設施以及其它公益與便利設施的距離進行衡量。關聯性的好處包括使更多基地使用者滿意，以及減少與使用公益設施相關的出行量。
Compact fluorescent lamp (CFL): a small fluorescent lamp, used as a more efficient alternative to incandescent lighting; also called a PL, twin-tube, or biax lamp (EPA).	緊湊型螢光燈（**CFL**）：小型螢光燈，作為更高效的白熾燈照明替代燈具；也稱之為光致發光、雙管或雙軸燈（美國環保局）。
Construction and demolition: debris waste and recyclables generated from construction and from the renovation, demolition, or deconstruction of preexisting structures. It does not include land- clearing debris, such as soil, vegetation, and rocks.	建造與拆除：從建造以及對已有結構的翻新、拆除或全面拆毀的過程中產生的廢棄雜物和可回收材料，不包括基地清理的雜物，如土壤、植被與岩石。
Construction waste management plan: a plan that diverts construction debris from landfills through recycling, salvaging, and reusing.	建築廢棄物管理計畫：通過再循環、回收及再利用來避免建築廢料被垃圾填埋的計畫。
Contaminant: an unwanted airborne element that may reduce indoor air quality (ASHRAE Standard 62.1-2007).	污染物：可降低室內空氣品質的有害空氣成分（ASHRAE 標準 62.1-2007）。
Controllability of systems: the percentage of occupants who have direct control over temperature, airflow, and lighting in their spaces.	系統可控性：可直接控制所在空間內溫度、氣流與照明的建築用戶百分比。

Cooling tower: a structure that uses water to absorb heat from air-conditioning systems and regulate air temperature in a facility.	冷卻塔：使用水來吸收空調系統產生的熱量並調節設施內氣溫的結構體。
Cradle to cradle: an approach in which all things are applied to a new use at the end of a useful life.	從搖籃到搖籃：將所有事物在使用壽命結束時投入新用途的處理方式。
Cradle to grave: a linear set of processes that lead to the ultimate disposal of materials at the end of a useful life.	從搖籃到墳墓：材料從被開始使用到使用壽命結束而被最終丟棄的線性過程。
Daylighting: the controlled admission of natural light into a space, used to reduce or eliminate electric lighting.	自然採光：將自然光有控制地調入空間，以減少或取消電氣照明。
Development density: the total square footage of all buildings within a particular area, measured in square feet per acre or units per acre.	開發密度：特定區域內所有建築物的總占地面積，以平方英尺／英畝或單位／英畝衡量。
Diversion rate: the percentage of waste materials diverted from traditional disposal methods to be recycled, composted, or re-used.	轉移率：由傳統處置方法轉為再循環、施堆肥或再利用的廢棄材料所占的百分比。
Diversity of uses or housing types: the number of types of spaces or housing types per acre. A neighborhood that includes a diversity of uses—offices, homes, schools, parks, stores—encourages walking, and its residents and visitors are less dependent on personal vehicles. A diversity of housing types allows households of different types, sizes, ages, and incomes to live in the same neighborhood.	用途多樣性或房屋類別：每英畝範圍內空間類型或房屋類別的數量。包括多種使用功能（如辦公室、住宅、學校、公園及商店）的社區鼓勵步行，其居民和訪客對私家車的依賴性更低。房屋類別的多樣性使不同類型、大小、年齡和收入的家庭得以在同一社區內居住。

Dry pond: an excavated area that detains stormwater and slow runoff but is dry between rain events. Wet ponds serve a similar function but are designed to hold water all the time.	乾池：保留雨水並減緩徑流，但在降雨之間保持乾涸的開挖區。濕池具有類似的功能，但設計為池內始終保留有水。
Ecosystem: a basic unit of nature that includes a community of organisms and their nonliving environment linked by biological, chemical and physical process.	生態系統：包括通過生物、化學及物理過程相聯結的生物群落及其非生物環境的基本自然單位。
Embodied energy: the total amount of energy used to harvest or extract, manufacture, transport, install and use a product across its life cycle.	蘊藏能量：在整個產品的生命週期中收穫或提取、製造、運輸、安裝及使用所消耗的能源總量。
Emergent properties: patterns that emerge from a system as a whole, which are more than the sum of the parts.	突現屬性：從一個系統中整體突現出的、大於各部分之和的特性。
Energy efficiency: using less energy to do the same amount of work.	節能：使用較少的能源完成同等的工作量。
Energy management system: a control system capable of monitoring environmental and system loads and adjusting HVAC operations accordingly in order to conserve energy while maintaining comfort (EPA).	能源管理系統：一種可監測環境與系統負荷，並相應調整暖通空調（HVAC）運行，以便在保持舒適性的同時節約能源的控制系統（美國環保局 EPA）。
ENERGY STAR Portfolio Manager: an interactive, online management tool that supports tracking and assessment of energy and water consumption.	能源之星資料管理器：支持能源與水消耗量跟蹤及評估的互動式線上管理工具。

ENERGY STAR rating: a measure of a building's energy performance compared with that of similar buildings, as determined by the ENERGY STAR Portfolio Manager. A score of 50 represents average building performance.	能源之星等級：由能源之星資料管理器確定的建築能源性能與類似建築的對比度量。50 分代表平均建築性能。
Energy use intensity: energy consumption divided by the number of square feet in a building, often expressed as British thermal units (Btus) per square foot or as kilowatt-hours of electricity per square foot per year (kWh/sf/yr).	能耗密度：能耗量除以建築面積（平方英尺），通常表示為英制熱量單位（Btus）／平方英尺或千瓦時／平方英尺／年（kWh/sf/yr）。
Energy-efficient products and systems: building components and appliances that use less energy to perform as well as or better than standard products.	節能產品與系統：使用較少的能源以實現相當於或優於標準產品性能的建築構件與設備。
Environmental sustainability: long-term maintenance of ecosystem components and functions for future generations (EPA).	環境永續性：生態系統要素與功能的長期維護，以造福子孫後代（美國環保局）。
Externality: costs or benefits, separate from prices, resulting from a transaction and incurred by parties not involved in the transaction.	外部因素：作為交易後果的、由未參與交易的各方引起的、獨立於價格以外的費用或利益。
Feedback loop: information flows within a system that allow the system to self-organize.	反饋回路：使系統實現自我組織的系統內部資訊流。
Floodplain: land that is likely to be flooded by a storm of a given size (e.g., a 100-year storm).	溯原：可能被一定規模的暴雨（如百年一遇的暴雨）淹沒的地帶。

Floor-area ratio: the relationship between the total building floor area and the allowable land area the building can cover. In green building, the objective is to build up rather than out because a smaller footprint means less disruption of the existing or created landscape.	容積率：總建築面積與建築所允許的占地面積的比率。由於較小的占地面積意味著對現有或修建景觀的較少破壞，綠色建築的目標是向上延伸建造而非向外擴張建造。
Flush-out: the operation of mechanical systems for a minimum of two weeks using 100 percent outside air at the end of construction and prior to building occupancy to ensure safe indoor air quality.	吹洗（大量換氣）：通過機械系統在施工結束後及建築投入使用前，採用100%室外空氣對建築至少吹洗兩周，以確保安全的室內空氣品質。
Footcandle: a measure of the amount of illumination falling on a surface. A footcandle is equal to one lumen per square foot. Minimizing the number of footcandles of site lighting helps reduce light pollution and protect dark skies and nocturnal animals.	尺燭光：落在一個表面上的光通量的量度。一尺燭光等於1流明／平方英尺。降低基地照明的尺燭光數有助於減少光污染，保護夜空及夜間活動的動物。
Fossil fuel: energy derived from ancient organic remains, such as peat, coal, crude oil, and natural gas (EPA).	化石燃料：源于古時有機殘餘物如泥炭、煤、原油及天然氣的能源（美國環保局）。
Gallons per flush (gpf): the amount of water consumed by flush fixtures (water closets, or toilets, and urinals).	每次沖水加侖數（**gpf**）：沖洗裝置（座便器或抽水馬桶、小便器）的用水量。
Gallons per minute (gpm): the amount of water consumed by flow fixtures (lavatory faucets, showerheads, aerators).	每分鐘加侖數（**gpm**）：水流裝置（面盆水龍頭、噴淋頭及通風裝置）的用水量。

Green building: a process for achieving ever-higher levels of performance in the built environment that creates more vital communities, more healthful indoor and outdoor spaces, and stronger connections to nature. The green building movement strives to effect a permanent shift in prevailing design, planning, construction, and operations practices, resulting in lower-impact, more sustainable, and ultimately regenerative built environments.	綠色建築：實現建築環境更高性能水準的過程。此類建築環境營造更重要的社區、更健康的室內與室外空間、與大自然有更強的聯繫。綠色建築運動力求影響主流設計、規劃、建造與運營實踐的永久性轉變，從而實現影響較低、永續性更好、可最終再生的建築環境。
Green power: energy from renewable sources such as solar, wind, wave, biomass, geothermal power and several forms of hydroelectric power.	綠色電力：源于可再生資源（如太陽能、風能、波浪能、生物質能、地熱發電及多種形式水力發電）的能源。
Greenfield: a site that has never been developed for anything except agriculture.	綠地：除農業以外，從未被開發用於其他用途的基地。
Greenhouse gas emissions per capita: a community's total greenhouse gas emissions divided by the total number of residents.	溫室排放量／每人：社區溫室氣體總排放量除以居民總數。
Greenwashing: presenting misinformation to consumers to portray a product or policy as more environmentally friendly than it actually is.	漂綠：誇大產品或政策的實際環保特性，誤導消費者。
Graywater: domestic wastewater composed of wash water from kitchen, bathroom, and laundry sinks, tubs, and washers. (EPA) The Uniform Plumbing Code (UPC) defines graywater in its Appendix G, Gray Water Systems for Single-Family	灰水：包括廚房、浴室、洗衣水槽、浴缸及洗衣機沖洗水的生活廢水。（美國環保局 EPA）《統一給排水規範（UPC）》附錄 G-「獨戶住宅灰水系統」中將灰水定義為「尚未接觸廁衛垃圾的未經處理的生活廢水。灰水包括浴缸、

Dwellings, as "untreated household waste water which has not come into contact with toilet waste. Graywater includes used water from bathtubs, showers, bathroom wash basins, and water from clothes-washer and laundry tubs. It must not include waste water from kitchen sinks or dishwashers." The International Plumbing Code (IPC) defines graywater in its Appendix C, Gray Water Recycling Systems, as "waste water discharged from lavatories, bathtubs, showers, clothes washers and laundry sinks." Some states and local authorities allow kitchen sink wastewater to be included in graywater. Other differences with the UPC and IPC definitions can likely be found in state and local codes.

Project teams should comply with graywater definitions as established by the authority having jurisdiction in the project area.

淋浴間與浴室面盆洗滌用水以及洗衣機與洗衣池用水，但不包括廚房水槽或洗碗機廢水。」
《國際給排水規範（IPC）》附錄C-「灰水循環系統」中將灰水定義為「廁所、浴缸、淋浴間、洗衣機和洗衣水槽排放的廢水」。
某些州和地方當局還將廚房水槽廢水納入灰水範疇。各個州與地方法規中可能存在與 UPC 及 IPC 定義的其它差異。專案團隊應遵循專案所在地區的管轄當局確定的灰水定義。

Hardscape: paved or covered areas in which soil is no longer on the surface of the Earth, such as roadways or parking lots.

硬景觀：土壤不再處於地表的鋪築或覆蓋區域，如道路或停車場。

Harvested: rainwater precipitation captured and used for indoor needs, irrigation, or both.

採集：被收集以用於室內需求或灌溉（或兩者兼有）的降雨。

Heat island effect: the absorption of heat by hardscapes, such as dark, nonreflective pavement and buildings, and its radiation to surrounding areas. Particularly in urban areas, other sources may include vehicle exhaust, air-conditioners, and street

熱島效應：硬景觀（如黑色非反射路面與建築）對熱量的吸收及其對周邊地區的輻射。在城市地區尤其顯著，其他熱源還可能包括汽車尾氣、空調和街道設備；高層建築與狹窄街道氣流的減少加劇了這一效應。

equipment; reduced airflow from tall buildings and narrow streets exacerbates the effect.	
High-performance green building: a structure designed to conserve water and energy; use space, materials, and resources efficiently; minimize construction waste; and create a healthful indoor environment.	高效能綠色建築：旨在節約水與能源的構築物；高效使用空間、材料和資源；最大程度地減少建築垃圾並營造健康的室內環境。
Hydrochlorofluorocarbon (HCFC): an organic chemical compound known to have ozone-depleting potential.	氫氯氟碳化合物（HCFC）：一種已知的、具備臭氧消耗潛能的有機化合物。
HVAC system: equipment, distribution systems, and terminals that provide the processes of heating, ventilating, or air-conditioning (ASHRAE Standard 90.1–2007).	冷凍空調系統：提供採暖、通風或空氣調節的設備、分配系統與終端（ASHRAE 標準 90.1-2007）。
Impervious area: surface that has been compacted or covered by materials that do not allow water to infiltrate. Impervious areas found in the built environment include concrete, brick, stone, asphalt, and sealed surfaces.	不透水區：已被壓實或覆蓋不透水材料的表面。建築環境中存在的不透水區包括混凝土、磚、石、瀝青和密封表面。
Imperviousness: the resistance of a material to penetration by a liquid. The total imperviousness of a surface, such as paving, is expressed as a percentage of total land area that does not allow moisture penetration. Impervious surfaces prevent rainwater from infiltrating into the ground, thereby increasing runoff, reducing groundwater recharge, and degrading surface water quality.	不滲透性：材料對液體滲透的耐受性。表面（如鋪築面）的總體不滲透性以總占地面積中不會發生水分滲透區域所占的百分比表示。不透水表面阻止雨水滲透進入地表內，導致徑流增加、地下水補給減少和地表水質下降。

Indoor air quality: the nature of air inside the space that affects the health and well-being of building occupants. It is considered acceptable when there are no known contaminants at harmful concentrations and a substantial majority (80% or more) of the occupants do not express dissatisfaction (ASHRAE Standard 62.1-2007).	室內空氣品質：影響建築用戶健康及福祉的室內空氣特性。如果已知的污染物未達到有害的濃度值，且絕大多數（80% 及以上）住戶未表達不滿時，室內空氣品質應視為可接受（ASHRAE 標準 62.1-2007）。
Indoor Air Quality Building Education and Assessment Model (I-BEAM): an integral part of an IAQ management program that provides comprehensive guidance for building professionals responsible for indoor air quality in commercial buildings. Incorporates an IAQ audit of the project building to determine the building's IAQ status.	建築室內空氣品質教育及評估模式（I-BEAM）：是為負責商業樓宇室內空氣品質的建築專業人士提供全面指導的室內空氣品質（IAQ）管理計畫中不可分割的一部分。結合專案建設的 IAQ 審核以確定建築的 IAQ 狀況。
Indoor environmental quality: the conditions inside a building and their impacts on occupants or residents.	室內環境品質：建築內部環境狀況及其對用戶或住戶的影響。
Indoor environmental quality management plan: a plan that spells out strategies to protect the quality of indoor air for workers and occupants; it includes isolating work areas to prevent contamination of occupied spaces, timing construction activities to minimize exposure to off-gassing, protecting the HVAC system from dust, selecting materials with minimal levels of toxicity, and thoroughly ventilating the building before occupancy.	室內環境品質管制計畫：一項闡明了工作人員及住戶室內空氣品質保護策略的計畫，包括隔離工作區防止已使用的空間受到汙染；安排施工活動的時間，盡可能減少接觸廢氣；HVAC 系統防塵；選用毒性最低的材料並於入住前對建築進行徹底通風。

Infill development: is a method of site selection that focuses construction on sites that have been previously developed or are gaps between existing structures.	填充式開發：一種將建築集中于已開發的基地或現有構築物之間的間隙基地上的選址方法。
Integrated approach: bringing team members together to work collaboratively on all of a project's systems to find synergistic solutions that support ever greater levels of sustainability.	整合方法：使團隊成員在專案所有系統上合作開展相關工作，以尋求支持更高水準永續性的協同解決方案。
Integrated design team: all the individuals involved in a building project from early in the design process, including the design professionals, the owner's representatives, and the general contractor and subcontractors.	整合設計團隊：從設計初期開始參與建設專案的所有個人，包括設計專業人員、業主代表以及總承包商與分包商。
Integrated pest management: a sustainable approach that combines knowledge about pests, the environment, and pest prevention and control methods to minimize pest infestation and damage in an economical way while minimizing hazards to people, property, and the environment.	害蟲綜合治理：一種將病蟲害和環境知識與害蟲防治及控制措施相結合的永續性方式，以經濟的方式盡可能減少害蟲侵襲與損害，同時最大限度地降低對人、財產和環境的危害。
Integrated process: an approach to design and operations that brings team members together to work collaboratively and find synergistic solutions that support ever greater levels of sustainability.	整合流程：一種使團隊成員合作一致開展工作，並尋求支持更高水準永續性的協同解決方案的設計與運營方法。
Irrigation efficiency: the percentage of water delivered by irrigation equipment that is actually used for irrigation and does not evaporate, blow away, or fall on hardscape. For example, overhead spray sprinklers have lower irrigation efficiencies (65%) than drip systems (90%).	灌溉效率：由灌溉設備輸送的，實際用於灌溉的，未被蒸發、吹走或灑落于硬景觀上的水量百分比。例如，頂噴式噴霧灑水噴頭的灌溉效率（65%）低於滴灌系統（90%）。

Iterative process: circular and repetitive process that provides opportunities for setting goals and checking each idea against those goals.	反覆運算過程：為設定目標並針對這些目標檢查每個想法提供機會的循環和重複過程。
LEED credit: an optional LEED Green Building Rating System component whose achievement results in the earning of points toward certification.	LEED 得分點：LEED 綠色建築評估體系的可選組成部分。它的獲得有助於贏得認證所需總分。
LEED credit interpretation request: a formal USGBC process in which a project team experiencing difficulties in the application of a LEED prerequisite or credit can seek and receive clarification, issued as a credit interpretation ruling. Typically, difficulties arise when specific issues are not directly addressed by LEED reference guides or a conflict between credit requirements arises.	LEED 得分點解釋請求：美國綠色建築委員會（USGBC）的一種正式流程。在此流程中，在對於某個 LEED 先決條件或得分點的問題上遇到困難時，專案團隊可尋求並收到作為得分點解釋裁決而出具的澄清書。通常情況下，當具體問題不能直接根據 LEED 參考指南解決，或在得分點要求之間存在衝突時，就會遇到困難。
LEED intent: the primary goal of each prerequisite or credit.	LEED 目的：每個先決條件或得分點的首要目標。
LEED Interpretation: consensus-based, precedent-setting rulings to project team inquiries.	LEED 解釋：對專案團隊所提出的問詢基於共識的先例裁決。
LEED Online: a data collection portal managed by GBCI through which the team uploads information about the project.	LEED 線上：由綠色建築認證協會（GBCI）管理的資料收集入口。團隊可通過此入口上傳專案資訊。
LEED Pilot Credit Library: credits currently being tested across rating systems and credit categories that are proposed for the next version of LEED.	LEED 試行得分點庫：在各個評估體系和分數類別中目前正在測試的得分點，以及為下一版 LEED 擬定的得分點類別。

LEED prerequisite: a required LEED Green Building Rating System component whose achievement is mandatory and does not earn any points.	**LEED 先決條件**：LEED 綠色建築評估體系所要求的必備組成部分。它們的實現是強制性的且不獲得任何得分。
LEED Rating System: a voluntary, consensus-based, market-driven building rating system based on existing, proven technology. The LEED Green Building Rating System represents USGBC's effort to provide a national benchmark for green buildings. Through its use as a design guideline and third-party certification tool, the LEED Green Building Rating System aims to improve occupant well-being, environmental performance, and economic returns using established and innovative practices, standards, and technologies.	**LEED 評估體系**：以共識為基礎，以市場為導向，基於現有成熟技術的自願性建築評估體系。LEED 綠色建築評估體系代表了 USGBC 在提供綠色建築國家基準方面所做出的努力。作為設計準則和第三方認證工具，LEED 綠色建築評估體系旨在通過採用既定的創新做法、標準和技術，以提高建築用戶的福祉、環保性能及經濟回報。
LEED technical advisory group (TAG): a committee consisting of industry experts who assist in interpreting credits and developing technical improvements to the LEED Green Building Rating System.	**LEED 技術諮詢顧問小組（TAG）**：由業內專家組成的委員會。目的是協助解釋 LEED 綠色建築評估體系得分點並對評估體系提出技術改進。
Leverage point: a point in a system where a small intervention can yield large changes.	**杠杆點**：系統中的某一點，該點處微小的干預即可產生很大的變化。
Life-cycle approach: looking at all stages of a project, product or service, adding the dimension of longevity to whole systems thinking.	**生命週期方法**：著眼於專案、產品或服務的所有階段，在整個系統考量中加入延長使用壽命的因素。
Life-cycle assessment: an analysis of the environmental aspects and potential impacts associated with a product, process, or service.	**生命週期評估**：對與產品、流程或服務有關的環境因素及其潛在影響的分析。

Life-cycle costing: a process of costing that looks at both purchase and operating costs as well as relative savings over the life of the building or product.	生命週期成本計算：著眼於建築或產品整個生命週期內採購與運營成本，以及相關成本節約的成本計算過程。
Light trespass: the spillage of light beyond the project boundary.	光侵擾：超過專案界區的光線溢散。
Lighting power density: the installed lighting power per unit area.	照明功率密度：單位面積內安裝的照明功率。
Low impact development (LID): an approach to land management that mimics natural systems to manage stormwater as close to the source as possible.	低衝擊開發（**LID**）：一種類比自然系統，盡可能接近水源進行雨水管理的土地管理方法。
Market transformation: systematic improvements in the performance of a market or market segment. For example, EPA's ENERGY STAR program has shifted the performance of homes, buildings, and appliances toward higher levels of energy efficiency by providing recognition and comparative performance information through its ENERGY STAR labels.	市場轉型：市場或細分市場表現的系統性改善。例如，通過提供能源之星標籤的識別與性能比較資訊，美國環保局的能源之星計劃已將住宅、建築與設備的性能向更高的能源效率轉變。
Materials reuse: materials returned to active use (in the same or a related capacity as their original use), expressed as a percentage of the total materials cost of a building. The salvaged materials are incorporated into the new building, thereby extending the lifetime of materials that would otherwise be discarded.	材料再利用：有效再利用的材料（與其原有用途相同，或具備與原有用途相關的性能），以占建築總材料成本的百分比表示。廢棄材料被納入新的建築中，從而延長了擬丟棄材料的使用壽命。

Measures of energy use: typical primary measures of energy consumption associated with buildings include kilowatt-hours of electricity, therms of natural gas, and gallons of liquid fuel.	能源使用度量：與建築相關的能源消耗典型主要度量單位，包括千瓦時（電）、克卡（天然氣）及加侖（液態燃料）。
Minimum Efficiency Reporting Value (MERV): a rating that indicates the efficiency of air filters in the mechanical system. MERV ratings range from 1 (very low efficiency) to 16 (very high).	最低效率報告值（**MERV**）：表明機械系統中空氣篩檢程式效率的一項評分值。MERV 分值介於 1（效率很低）至 16（效率很高）之間。
Montreal Protocol: an international treaty that eliminates or partially eliminates the use of substances known to deplete the ozone layer by phasing out production of substances such as CFCs and HCFCs.	蒙特婁議定書：一項旨在通過逐步淘汰氯氟氟碳及氫氯氟碳化合物等物質的生產，以消除或部分消除使用已知消耗臭氧層的物質的國際公約。
native (or indigenous) plant: a plant adapted to a given area during a defined time period; in North America, the term often refers to plants growing in a region prior to the time of settlement by people of European descent. Native plants are considered low maintenance and not invasive.	本地（或本土）植物：一種在規定的時段內適應某指定地區的植物；在北美，本術語通常是指在歐洲血統的人定居之前某一地區內生長的植物。本地植物的維護成本低且不具入侵性。
Negative feedback loop: a signal for a system to stop changing when a response is no longer needed.	負反饋循環：當不再需要回應時，使系統停止更改的信號。
Nonpoint source pollution: typically refers to water pollution caused by stormwater runoff from diffuse sources. When it rains, water washes fertilizers, car oil, pet waste, etc, into receiving water bodies.	非點源污染：通常是指由從擴散源徑流的雨水引起的水質污染。下雨時，雨水將化肥、車油、寵物糞便等沖刷至受納水體內。

Nonpotable water: See potable water.	非自來水：請參見自來水。
Nonrenewable: not capable of being replaced; permanently depleted once used. Examples of nonrenewable energy sources are oil or natural gas, and nonrenewable natural resources include metallic ores.	不可再生：不可恢復的；一經使用即永久耗盡的。不可再生能源的實例為石油或天然氣；不可再生的自然資源包括金屬礦石。
Occupant comfort survey: measures occupant comfort level in a variety of ways, including thermal comfort, acoustics, indoor air quality, lighting levels, and building cleanliness.	住戶舒適性調查：以各種方式衡量住戶的舒適程度，包括熱舒適性、聲學、室內空氣品質、照明水準及建築清潔度。
Off-gassing: the emission of volatile organic compounds from synthetic and natural products.	廢氣排放：人造與天然產品的揮發性有機化合物排放量。
Open system: a system in which materials are constantly brought in from the outside, used in the system, and then released outside the system in a form of waste.	開放系統：一種不斷從外界引入材料並在系統內使用，然後將材料以廢棄物形式排出的系統。
Particulates: a solid particle or liquid droplets in the atmosphere; the chemical composition of particulates varies, depending on location and time of year. Sources include dust, emissions from industrial processes, combustion products from the burning of wood and coal, combustion products associated with motor vehicle or nonroad engine exhausts, and reactions to gases in the atmosphere (EPA).	微粒：大氣中的固體顆粒或液滴；根據位置及每年時間的不同，微粒的化學成分會有所差異。微粒源包括粉塵、工業過程排放物、木材與煤炭的燃燒產物、機動車或非道路發動機排氣以及大氣中與燃氣反應相關的產物（美國環保局 EPA）。

Passive design: planning with the intent of capturing natural elements such as sunlight and wind for light, heating, and cooling.	被動式設計：旨在通過獲取自然元素（如陽光與風能）來解決照明、採暖與製冷的設計。
Performance monitoring: continuously tracking metrics of energy, water and other systems, specifically to respond and achieve better levels of efficiency.	性能監控：連續跟蹤能源、水及其它系統的指標，專門用於回應並實現更高的效率。
Performance relative to benchmark: a comparison of the performance of a building system with a standard, such as ENERGY STAR Portfolio Manager.	相對於基準的性能：建築系統性能與現行標準（如能源之星資料管理器）的比較。
Performance relative to code: a comparison of the performance of a building system with a baseline equivalent to minimal compliance with an applicable energy code, such as ASHRAE Standard 90.1 or California's Title 24.	相對於規範的性能：建築系統性能與相當於適用最低能源規範（如 ASHRAE 標準 90.1 或加州法規第 24 章）的基線的相比較。
Perviousness: the percentage of the surface area of a paving material that is open and allows moisture to pass through the material and soak into the ground below.	透水性：開敞的，以及允許水分通過材料滲透入地面以下的鋪築材料表面積所占百分比。
Pest control management: a sustainable approach that combines knowledge about pests, the environment, and pest prevention and control methods to minimize pest infestation and damage in an economical way while minimizing hazards to people, property, and the environment.	害蟲控制管理：一種將病蟲害、環境知識與害蟲防治及控制措施相結合的永續性方式，以經濟的方式盡可能減少害蟲侵襲與損害，同時最大限度地降低對人、財產和環境的危害。

Photovoltaic (PV) energy: electricity from photovoltaic cells that convert the energy in sunlight into electricity.	太陽光電板（**PV**）能源：將太陽能轉化為電能的，由光電電池所產生的電力。
Pollutant: any substance introduced into the environment that adversely affects the usefulness of a resource or the health of humans, animals, or ecosystems. (EPA) Air pollutants include emissions of carbon dioxide (CO_2), sulfur dioxide (SO_2), nitrogen oxides (NOx), mercury (Hg), small particulates (PM2.5), and large particulates (PM10).	污染物：任何一種引入環境中，並對資源的可用性或人類健康、動物及生態系統產生不利影響的物質。（美國環保局）空氣污染物包括二氧化碳（CO_2）、二氧化硫（SO_2）、氮氧化物（NOx）、汞（Hg）、細小微粒（PM2.5）及大顆粒物（PM10）的排放。
Positive feedback loop: self-reinforcing loops in which a stimulus causes an effect and the loop produces more of that effect.	正回饋循環：一種自我強化循環。其中的某種刺激引起某種效應時，該循環會產生更多的此種效應。
Postconsumer recycled content: the percentage of material in a product that was consumer waste. The recycled material was generated by household, commercial, industrial, or institutional end users and can no longer be used for its intended purpose. It includes returns of materials from the distribution chain. Examples include construction and demolition debris, materials collected through recycling programs, discarded products (e.g., furniture, cabinetry, decking), and landscaping waste (e.g., leaves, grass clippings, tree trimmings) (ISO 14021).	消費後回收物質含量：產品原料中所回收的在消費後產生的廢棄物所占的百分比。回收材料是來自於生活、商業、工業或機構最終用戶，且不可再用於其原有用途的材料。其中包括從分銷鏈回收的材料。例如：房屋建造與拆除產生的廢棄物、通過回收計畫收集的材料、棄用品（如傢俱、櫥櫃、裝飾）以及景觀廢棄物（如樹葉、草屑、樹木修剪碎屑）（ISO14021）。

Potable water: water that meets or exceeds EPA's drinking water quality standards and is approved for human consumption by the state or local authorities having jurisdiction; it may be supplied from wells or municipal water systems.	飲用水：滿足或超過美國環保局自來水水質標準，且經由州或地方管轄當局批准用於人類消費的水源；可通過水井或市政供水系統供給。
Preconsumer recycled content: the percentage of material in a product that was recycled from manufacturing waste. Preconsumer content was formerly known as postindustrial content. Examples include planer shavings, sawdust, bagasse, walnut shells, culls, trimmed materials, overissue publications, and obsolete inventories. Excluded are rework, regrind, or scrap materials capable of being reclaimed within the same process that generated them (ISO 14021).	消費前回收物質含量：產品中從製造過程中產生的廢棄物回收的材料產品中所占的百分比。消費前含量以前稱為工業用後含量。例如：刨花、鋸屑、甘蔗渣、核桃殼、下腳料、修剪材料、過量發行的出版物及陳舊存貨，但不包括可通過與生產材料相同的過程予以回收的返工、粉碎再生材料或廢料（ISO14021）。
Prime farmland: previously undeveloped land with soil suitable for cultivation. Avoiding development on prime farmland helps protect agricultural lands, which are needed for food production.	基本農田：土壤適合耕種且之前未經開發的土地。避免對基本農田的開發，有利於保護糧食生產所需的農業用地。
Project administrator: the individual from the project team that registers a project with GBCI.	專案管理員：在綠色建築認證協會（GBCI）註冊專案的專案團隊成員。
Project credit interpretation rulings (CIR): a response from GBCI providing technical guidance on how LEED requirements pertain to particular projects.	專案得分點解釋裁決（CIR）：綠色建築認證協會（GBCI）就LEED要求如何適合於特定專案提供技術指導的答覆。

Project team: a broad, inclusive, collaborative group that works together to design and complete a project.	專案團隊：一個共同設計並完成專案的，具有廣泛包容性的協作組團。
Rain garden: a stormwater management feature consisting of an excavated depression and vegetation that collect and infiltrate runoff and reduce peak discharge rates.	雨水花園：一種包括收集和滲透徑流，並降低洪峰流量的開挖窪地和植被的雨水管理設施特徵。
Rainwater harvesting: the collection and storage of precipitation from a catchment area, such as a roof.	雨水收集：從集水區（如屋面）收集並儲存降水。
Rapidly renewable materials: agricultural products (fiber or animal) that are grown or raised quickly and can be harvested in a sustainable fashion, expressed as a percentage of the total materials cost. For LEED, rapidly renewable materials take 10 years or less to grow or raise.	快速可再生材料：能夠快速生長或養殖，且可通過可持續的方式收穫的農產品（纖維或動物），以總材料成本百分比表示。對於LEED而言，快速可再生材料需要10年或以內的時間生長或養殖。
Recycled content: the percentage of material in a product that is recycled from the manufacturing waste stream (preconsumer waste) or the consumer waste stream (postconsumer waste) and used to make new materials. For LEED, recycled content is typically expressed as a percentage of the total material volume or weight.	回收物質含量：產品的原料中由從工業生產廢物流（消費前廢棄物）或消費使用後垃圾廢物流（消費後廢棄物）中回收的用以製造新材料所占的百分比。對於LEED而言，回收物質含量通常表示為總材料體積或重量百分比。
Refrigerant: one of any number of substances used in cooling systems to transfer thermal energy in air conditioning and refrigeration systems.	製冷劑：用於冷卻系統中轉移空調及製冷系統中熱能的任何物質之一。
Regenerative: evolving with living systems and contributing to the long term renewal of resources and the health of all life in each unique place.	再生性：隨生命系統而演變，並有助於資源的長期再生以及每一獨特地區內所有生物的健康。

Regenerative design: sustainable plans for built environments that improve existing conditions. Regenerative design goes beyond reducing impacts to create positive change in the local and global environment.	再生性設計：改善現有條件的建築環境永續性計畫。再生性設計不僅降低影響，還對當地及全球環境產生積極性的影響。
Regional material: a material that is extracted, processed, and manufactured close to a project site, expressed as a percentage of the total materials cost; for LEED, regional materials originate within 500 miles of the project site.	本地材料：於專案現場就近選取、加工及製造的材料，以占總材料成本的百分比表示；對於 LEED 而言，地域性材料源於專案現場 500 英里範圍內。
Renewable energy: resources that are not depleted by use. Examples include energy from the sun, wind, and small (low-impact) hydropower, plus geothermal energy and wave and tidal systems. Ways to capture energy from the sun include photovoltaic, solar thermal, and bioenergy systems based on wood waste, agricultural crops or residue, animal and other organic waste, or landfill gas.	可再生能源：不因使用而耗盡的資源。例如：太陽能、風能、小型（低影響）水力發電，以及地熱能與潮波系統。獲取太陽能的方法包括光伏、太陽熱能、基於木材廢料的生物能系統、農作物或殘渣、動物和其它有機廢棄物或垃圾場所產生的氣體。
Renewable energy certificate (REC): a tradable commodity representing proof that a unit of electricity was generated from a renewable energy resource. RECs are sold separately from the electricity itself and thus allow the purchase of green power by a user of conventionally generated electricity.	可再生能源認證（REC）：證明單位電量是源自可再生資源的一種可交易商品。REC 與電力本身分開而單獨出售，從而使傳統電力的用戶得以購買綠色電力。

Retrocommissioning: a commissioning process that can be performed on existing buildings to identify and recognize improvements that can improve performance.	重新功能驗證：可對現有建築實施的功能驗證過程，以識別並確認可提高建築性能的各項改進措施。
Salvaged material: construction items recovered from existing buildings or construction sites and reused. Common salvaged materials include structural beams and posts, flooring, doors, cabinetry, brick, and decorative items.	可重新利用的建築構件：從現有建築或施工現場回收並重複使用的建築構件。常見再利用材料包括結構樑柱、地板、門、櫥櫃、磚及裝飾構件。
Sick building syndrome (SBS): a combination of symptoms, experienced by occupants of a building, that appear to be linked to time spent in the building but cannot be traced to a specific cause. Complaints may be localized in a particular room or zone or be spread throughout the building (EPA).	病態建築綜合症（**SBS**）：建築住戶經歷的，可能與在建築內逗留時間有關但無法追查具體原因的綜合症狀。病症可能局限於特定的房間或區域內，或在整棟建築內傳播（美國環保局）。
Site disturbance: the amount of a site that is disturbed by construction activity. On undeveloped sites, limiting the amount and boundary of site disturbance can protect surrounding habitat.	現場干擾：現場受到施工活動干擾的程度。在未開發的基地，限制現場干擾的程度與範圍可保護周圍的生活環境。
Smart growth: an approach to growth that protects open space and farmland by emphasizing development with housing and transportation choices near jobs, shops and schools.	智慧型成長：通過強調開發與就業機會、商店和學校相鄰的住房與交通方案，並以此保護開放空間及農田的一種增長方式。
Solar reflectivity index (SRI): a measure of how well a material rejects solar heat; the index ranges from 0 (least reflective) to 100 (most	太陽能反射指數（**SRI**）：衡量材料抗拒太陽熱能程度的指標；指數介於0（反射率最低）至100（反射率最高）之間。使用淺色的「冷

reflective). Using light-colored, "cooler" materials helps prevent the urban heat island effect (the absorption of heat by dark roofs and pavement and its radiation to the ambient air) and minimizes demand for cooling of nearby buildings.	性」材料有助於防止城市熱島效應（深色屋面和路面吸收熱量並對周圍空氣產生輻射），並最大程度地降低附近建築的冷負荷。
Stakeholder: a dynamic term that encompasses a broad array of individuals tasked with the design, creation, and operation of a building as well as those whose lives will be impacted by the built environment at hand.	利益相關者：動態術語，包括承擔建築設計、建設及運營的各類人員，以及生活會受到附近建築環境影響的人員。
Stakeholder meeting: a meeting that includes those with a vested interest in the outcome of a project.	利益相關者會議：包括專案成果既得利益者出席的會議。
Stormwater prevention plan: a plan that addresses measures to prevent erosion, sedimentation, and discharges of potential pollutants to water bodies and wetlands.	逕流保護計畫：一項為水土流失、沉積，以及潛在污染物排放至水體或濕地中提供預防措施的計畫。
Stormwater runoff: water from precipitation that flows over surfaces into sewer systems or receiving water bodies. All precipitation that leaves project site boundaries on the surface is considered stormwater runoff.	雨水徑流：經地表流入污水系統或受納水體的降水。所有流出項目用地範圍的地表降水均視為雨水徑流。
Street grid density: an indicator of neighborhood density, calculated as the number of centerline miles per square mile. Centerline miles are the length of a road down its center. A community with high street grid density and narrow, interconnected streets is more likely to be pedestrian friendly than one with a low street grid density and wide streets.	街道網格密度：社區密度指標，以每平方英里的中軸英里數計算。中軸英里為道路在中軸線的長度。與街道網格密度低及街道寬闊的社區相比，街道網格密度高及互聯街道狹窄的社區更方便行人。

Sustainability: meeting the needs of the present without compromising the ability of future generations to meet their own needs (Brundtland Commission).	永續性：滿足當代人的需求又不損害子孫後代滿足自身需求的能力（布倫特蘭委員會）。
Sustainable forestry: management of forest resources to meet the long-term forest product needs of humans while maintaining the biodiversity of forested landscapes. The primary goal is to restore, enhance, and sustain a full range of forest values, including economic, social, and ecological considerations.	永續林業：管理森林資源以滿足人類的長期林業產品需求，同時保持森林景觀的生物多樣性。首要的目標是恢復、加強及維護全方位的森林價值，包括經濟、社會及生態因素。
Sustained-yield forestry: management of a forest to produce in perpetuity a high-level annual or regular periodic output, through a balance between increment and cutting (Society of American Foresters).	持續產出林業：通過增加與削減之間的平衡來管理森林，以永久保持較高的林業產品年產出量或定期產出量（美國森林學會）。
System: an assemblage of parts that interact in a series of relationships to form a complex whole, which serves particular functions or purposes.	系統：各部分通過一系列相互作用而形成的複雜整體，該整體提供了特定的功能或作用。
Systems thinking: understanding the built environment as a series of relationships in which all parts influence many other parts.	系統思想：將建築環境理解為一系列關係的組合的思想方式。在這一組合中，所有部分均會影響許多其他部分。
Task group: a small group organized to dive deeper into proposed ideas; groups investigate and refine proposed strategies	工作組：一個深層次研究擬議想法的小組；各小組調查並改進擬定的策略。
Team meetings: a meeting that follows the charrettes and continues throughout the iterative process; fostering collaboration and encouraging	團隊會議：在專家研討會後以及整個反覆運算過程中持續召開的會議；促進合作並鼓勵創造力，為探討如何使戰略適應整個專案

creativity, they provide opportunities to explore how strategies fit together in the context of the whole project.	提供機會。
Thermal comfort: the temperature, humidity, and airflow ranges within which the majority of people are most comfortable, as determined by ASHRAE Standard 55. Because people dress differently depending on the season, thermal comfort levels vary with the season. Control setpoints for HVAC systems should vary accordingly to ensure that occupants are comfortable and energy is conserved.	熱舒適：由 ASHRAE 標準 55 確定的，大多數人感覺舒適的溫度、濕度與氣流範圍。由於人們的著裝因季節而異，熱舒適度隨季節而變化。暖通空調系統的控制設定值應相應變化，以確保住戶的舒適性且節約能源。
Transportation demand management: the process of reducing peak-period vehicle trips.	交通需求管理：減少高峰期車輛出行的過程。
Triple bottom line: incorporates a long-term view for assessing potential effects and best practices for three kinds of resources: people, planet, profit.	**Triple Bottom Line**：評估潛在影響以及三類資源（人力、地球、利潤）最佳實踐的戰略遠景。
Value engineering: a formal review process of the design of a project based on its intended function in order to identify potential alternatives that reduce costs and improve performance.	價值工程：基於項目的預期功能對專案設計進行的正式評審過程，目的是確定可降低成本並改善性能的潛在替代品。
Vehicle miles traveled (vmt): a measure of transportation demand that estimates the travel miles associated with a project, most often for single-passenger cars. LEED sometimes uses a complementary metric for alternative-mode miles (e.g., in high-occupancy autos).	車輛行駛里程（**vmt**）：一個專案相關的行駛里程的交通需求估量。最常用于單乘客車輛。LEED 有時針對替代模式里程使用互補性度量（如高乘載率汽車）。

Ventilation rate: the amount of air circulated through a space, measured in air changes per hour (the quantity of infiltration air in cubic feet per minute divided by the volume of the room). Proper ventilation rates, as prescribed by ASHRAE Standard 62, ensure that enough air is supplied for the number of occupants to prevent accumulation of carbon dioxide and other pollutants in the space.	通風率：以每小時的換氣次數衡量（空氣滲透量（立方英尺／分鐘）除以房間的容量）的通過空間的氣流量。ASHRAE 標準 62 規定的適當通風率可確保為各住戶提供足夠的空氣，以防止二氧化碳和其它污染物在空間內的累積。
Volatile organic compound (VOC): a carbon compound that participates in atmospheric photochemical reactions (excluding carbon monoxide, carbon dioxide, carbonic acid, metallic carbides and carbonates, and ammonium carbonate). Such compounds vaporize (become a gas) at normal room temperatures. VOCs off-gas from many materials, including adhesives, sealants, paints, carpets, and particle board. Limiting VOC concentrations protects the health of both construction personnel and building occupants.	揮發性有機化合物（**VOC**）：一種參與大氣光化學反應的碳化合物（不包括一氧化碳、二氧化碳、碳酸、金屬碳化物、碳酸鹽及碳酸銨）。此類化合物在常溫條件下可蒸發成氣體。VOC 揮發氣體可產生於眾多材料，包括粘合劑、密封劑、塗料、地毯和刨花板。限制 VOC 濃度可保護施工人員及建築住戶的健康。
Waste diversion: the amount of waste disposed other than through incineration or in landfills, expressed in tons. Examples of waste diversion include reuse and recycling.	廢棄物轉移：通過除了焚燒或填埋以外的方式處置的廢棄物量，以噸表示。廢棄物轉移的實例包括再利用和再循環。
Waste management plan: a plan that addresses the sorting, collection, and disposal of waste generated during construction or renovation. It must address management of landfill waste as well as recyclable materials.	廢棄物管理計畫：一項解決施工或改建期間產生的廢棄物的分類、收集與處置問題的計畫。該計畫必須闡述對填埋廢棄物與可回收材料的管理。

Wastewater: the spent or used water from a home, community, farm, or industry that contains dissolved or suspended matter. (EPA)	廢水：家庭、社區、農場或工業產生的或使用的，含有可溶物或懸浮物的水（美國環保局）。
Water balance: a maximum for how much water tenants can use equal to the amount of rainfall that falls naturally on a site per year.	水平衡：租戶可使用的最高耗水量，等同於現場每年的自然降雨量。
Wetland vegetation: plants that require saturated soils to survive or can tolerate prolonged wet soil conditions.	濕地植物：需要飽和土壤才能生存或可長期耐受濕潤土壤條件的植物。
Wingspread Principles on a U.S. Response to Global Warming: a set of propositions signed by individuals and organizations declaring their commitment to addressing the issue of climate change.	美國應對全球變暖的處理原則：一套由個人和組織簽署的，宣稱致力於解決氣候變化問題的主張。
Xeriscaping: a landscaping method that makes routine irrigation unnecessary by using drought-adaptable and low-water plants, as well as soil amendments such as compost and mulches to reduce evaporation.	旱生園藝：通過採用耐旱和低耗水植物以及使用土壤改良物如堆肥和覆層來減少蒸發，無需進行常規灌溉的一種綠化方法。

12.4 LEED Green Associate 報考流程與注意事項

報名參加考試

註冊

1. 到官方 USGBC® 官網註冊一個 Credentials 帳戶。

2. 檢查您輸入的姓名是否與您將在考試中心出示的身分證明上的姓名匹配（需與護照上相同）。

3. 您將被導向至 prometric.com/gbci 來安排考試日期和地點。

4. 完成考試預約後,您可在螢幕上看到確認代碼,並通過電子郵件收到 Prometric 發送的此代碼。

5. 請記錄您的確認碼。您需要此確認碼在 Prometric 網站:prometric.com/gbci 確認、取消或重新安排預約。

6. 一旦考試時間確定,請將 Prometric 網站上您的確認通知列印出來。請妥善保存您的確認通知,以便與 Prometric 網站就考試進行溝通。

考試費用

LEED Green Associate 考試費用爲 250 美金(線上信用卡繳費),學生可以享有優惠折扣(可能需要提供學生證明)。

考試介紹

考試題目類別

- 記憶題:這些題目評估考生記憶在考試參考資料中規範或概念的能力。

- 應用題:這些題目給考生提出假想的問題或情景,考生可運用考試參考資料中所述的、熟知的原則或程式予以解決。

- 分析題:這些題目評估考生將問題分解爲各個組成部分並建立解決方案的能力。考生不僅應認識到形成問題的不同因素,還應評估這些因素之間的關係或相互作用。

考試形式

LEED® Green Associate™ 包括 100 道隨機提供的單複選題,且必須於 2 小時內完成。

- 單場考試包括 10 分鐘的介紹(可跳過)、2 小時的考試(Green

Associate 或專業考試）和 10 分鐘的退出調查問卷（可跳過）。

考試語種

考試的主要語言爲英語。目前也提供簡體中文的選項供輔助語言（需在註冊考試時選取 Chinese）。

考前備忘錄

考試中心簡要情況介紹

建議您至少在預定的考試時間之前 30 分鐘到達考試中心。於預定的考試時間之後到達考場的考生將不允許進入考場。

台灣的考場爲台大語言測驗中心（LTTC），地址：106 台北市大安區辛亥路 2 段 170 號，電話：02 23626385。

身份證明要求

考生必須提供附有簽名、本人照片和在有效期內的有效身分證件。由於必須要有照片與英文姓名，因此建議爲護照。

考試開始前會考場人員會仔細檢查身上所有物品，並需將資料與物品置於置物櫃。

考試結束後

考試成績

所有 LEED® 專業考試的得分均介於 125 至 200 分之間。170 分或以上被視爲通過考試。您的考試分數將在考試結束時顯示於螢幕上，且您會收到考試中心工作人員提供的列印成績報告單。通過後證書將在三天內可以至 USGBC 官網下載電子證書。

通過考試

1. 授予認證

一旦您通過 Green Associate 考試，可立即使用 "LEED® Green Associate ™" 的名稱和／或標誌。

2. 證書

一旦您的考試成績得到處理，即可要求提供證書。證書有兩種形式：PDF 電子證書（可隨時免費下載）和紙本（需要額外至官網訂購並付費）。

3. 認證維持方案

通過 LEED Green Associate 之後每兩年需要取得 15 小時的繼續教育時數。

4. 繼成爲 **LEED Green Associate** 之後取得 **LEED AP** 證書

考試詳述

詳細內容

LEED Green Associate，測試您掌握的商業建築和住宅建築、新建建築和既有建築相關的綠色建築實踐的一般知識，以及了解如何爲負責 LEED 專案的其他專業人員提供支援。

任務領域（Green Associate）

LEED Green Associate 任務（100%）

- 與團隊或同事溝通廣泛和基本的綠色建築理念
- 研究並創建可持續建築材料庫
- 說明他人實現永續性目標
- 製作專案概況／案例研究／新聞稿
- 向客戶、團隊成員和公眾宣導綠色理念（如，爲什麼要建造綠色建築）

- 了解對 LEED 和一般綠色策略的任何更新
- 瀏覽 LEED 線上（LEED Online）的內容
- 幫助專案負責人管理與所有專案團隊成員（顧問、承建商、業主等）之間的 LEED 文件
- 說明管理記錄過程
- 幫助管理 LEED 認證的時間安排

知識領域

LEED® 流程（**16 個問題**）

- 組織基礎（如使命／願景、非營利、USGBC®／GBCI 的作用）
- LEED 評估體系的結構（如得分點類別、先決條件、LEED 認證的得分點和／或最低專案要求）
- 每個 LEED 評估體系的範圍（如評估體系選擇、評估體系系列（BD+C、ID+C、O+M、ND、住宅））
- LEED 開發流程（如基於共識、利益相關者與志願者參與、評估體系更新／演變）
- 得分點類別（如 LT、SS、WE、EA、MR、EQ、IN、RP 各項的目標和目的；協同）
- 影響類別（如，LEED 專案應實現哪些成就？）
- LEED 認證流程（如認證等級（認證級、銀級、金級和白金級）、LEED 得分卡、協力廠商認證、檔遞交的作用、LEED 解釋、附錄、了解不同系統版本（如 LEED 線上（LEED Online））和 LEED 線上（LEED Online）和專案註冊的組成部分）
- 其他評估體系（例如：一般來說還有其他哪些評估體系？）

整合策略（**8 個問題**）

- 整合過程（例如：對系統間相互關係的早期分析、系統思想、專家研討會）

- 整合專案團隊成員（例如：建築師、工程師、景觀建築師、土建工程師、承建商、物業經理等）
- LEED 的輔助標準（如對美國採暖、製冷與空調工程師學會（ASHRAE）的廣泛而不是深入的了解、金屬散熱與空調承包商協會（SMACNA）指南、Green Seal、ENERGY STAR®、HER、ACP 中列出的參考標注等）

位置與交通（**7** 個問題）

- 選址（例如：在先前開發過的區域和褐地／高優先順序指定區域選址、避免敏感的棲息地、在包含既有基礎設施和附近服務功能的地區選址、停車面積減量）
- 替代性交通（例如：類型、方便程度和品質，基礎設施和設計）

永續基地（**7** 個問題）

- 場址評估（例如：環境評估、人為影響）
- 場址設計和開發（例如：施工活動汙染預防、棲息地保護和恢復、外部開放空間、雨水管理、外部照明、減少熱島效應）

用水效率（**9** 個問題）

- 室外用水量（例如：灰水／雨水在灌溉中的使用、本地和可適應性物種的使用）
- 室內用水量（例如：低流量／無水設備的概念、節水器具、類型和品質）
- 用水效率管理（如，測量和監控）

能源與大氣（**10** 個問題）

- 建築負荷（例如：建築構件、空間使用（私人辦公室、個人空間、共用多住戶空間））
- 能源效率（例如：設計基本理念、運營能源效率、調試、能源審計）

- 替代性和可再生能源實踐（例如：需求回應、可再生能源、綠色
  電力、碳補償）
- 能源表現管理（例如：能源使用測量和監控、建築自動化控制／
  高級能源計量、運營與管理、基準 測試、ENERGY STAR®）
- 環境問題（例如：資源和能源、溫室氣體、全球變暖潛能值、資
  源枯竭、臭氧消耗）

材料與資源（**9** 個問題）

- 再利用（例如：建築再利用、材料再利用、室內再利用、傢俱再
  利用）
- 生命週期影響（例如：生命週期評估的概念、材料屬性、人類與
  生態健康影響、設計的靈活性）
- 廢棄物（例如：建造與拆除、維護與翻新、營運與持續、廢棄物
  管理計畫）
- 採購和聲明（例如：採購政策與計畫、環保採購（EPP）、建築產
  品分析公示和優化（即，原材料采 購、材料成分、環保產品分析
  公示））

室內環境品質（**8** 個問題）

- 室內空氣品質（例如：通風等級、菸害控制、室內空氣品質管制
  和改善、低逸散性材料、綠色清潔）
- 照明（例如：電氣照明品質、自然採光）
- 聲音（例如：聲環境）
- 住戶舒適度、健康和滿意度（例如：系統可控性、熱舒適設計、
  視野品質、評估／調查）

專案周邊環境和公共推廣（**11** 個問題）

- 建築環境的環境影響（例如：傳統建築中的能源和資源使用、綠
  色建築的必要性、環境外部因素、三重底線）

- 規範（例如：LEED® 和規範〔建築、管道、電氣、機械、消防〕之間的關係、綠色建築規範）
- 永續性設計的價值（例如：持續節能、改善住戶健康狀況、費用節約措施、成本（硬成本、軟成本）、生命週期）
- 本地設計（例如：本地綠色設計和適當的施工措施，本地重點應放在可持續場址和材料與資源中）

樣題

　　由於正式考試為簡體中文，因此以下特以簡體中文表示之：

LEED Green Associate 考題

1. 在申请创新得分点时，项目团队

    (A) 无法递交任何之前获得的创新得分点。

    (B) 会因翻倍实现得分点要求阈值的表现而得到得分点。

    (C) 可以递交现有 LEED® 得分点中使用的产品或策略。

    (D) 对於项目团队中的每个 LEED Accredited Professional，都可以得到一个得分点。

该问题考察知识领域 A. LEED 流程，得分点分类和任务领域 A. LEED Green Associate 任务

2. 某开发商希望通过建造一个拥有最大自然采光和视野的新办公室来盈利。该开发商应采取什麽措施来满足 triple bottom line 的所有要求？

    (A) 恢复场址内的栖息地

    (B) 采购符合人体工学的家俱

    (C) 争取地方津贴和奖励

    (D) 为住户提供照明可控性

该问题考察知识领域 I. 项目周边环境和推广，建成环境的环境影响，以及任务领域 A. LEED Green Associate 任务，说明他人实现可持续性目标

12.5 LEED 得分表（LEED Scorecard）（繁體中文版）

SCORECARD

LEED v5 for Design and Construction: New Construction

Category / Credit	Points	Decarbonization	Quality of Life	Ecosystem Conservation & Restoration
Integrative Process, Planning & Assessments	**1**			
Prereq Climate Resilience Assessment	Required		•	
Prereq Human Impact Assessment	Required		•	
Prereq Carbon Assessment	Required	•		
Credit Integrative Design Process	1	•	•	•
Location & Transportation	**15**			
Credit Sensitive Land Protection	1			•
Credit Equitable Development	2		•	
Credit Compact and Connected Development	6	•	•	•
Credit Transportation Demand Management	4	•	•	
Credit Electric Vehicles	2	•		
Sustainable Sites	**11**			
Prereq Minimize Site Disturbance	Required			•
Credit Biodiverse Habitat	2			•
Credit Accessible Outdoor Space	1		•	
Credit Rainwater Management	3			•
Credit Enhanced Resilient Site Design	2		•	•
Credit Heat Island Reduction	2	•		•
Credit Light Pollution Reduction	1			•
Water Efficiency	**9**			
Prereq Water Metering and Reporting	Required			•
Prereq Minimum Water Efficiency	Required	•		•
Credit Water Metering and Leak Detection	1	•	•	•
Credit Enhanced Water Efficiency	8	•		•
Energy & Atmosphere	**33**			
Prereq Operational Carbon Projection and Decarbonization Plan	Required	•		
Prereq Minimum Energy Efficiency	Required	•		
Prereq Fundamental Commissioning	Required	•		
Prereq Energy Metering and Reporting	Required	•		
Prereq Fundamental Refrigerant Management	Required	•		
Credit Electrification	5	•		
Credit Reduce Peak Thermal Loads	5	•		
Credit Enhanced Energy Efficiency	10	•		
Credit Renewable Energy	5	•		
Credit Enhanced Commissioning	4	•		
Credit Grid Interactive	2	•		
Credit Enhanced Refrigerant Management	2	•		
Materials & Resources	**18**			
Prereq Planning for Zero Waste Operations	Required	•		
Prereq Assess and Quantify Embodied Carbon	Required	•		
Credit Building and Materials Reuse	3	•		
Credit Reduce Embodied Carbon	6	•		
Credit Low-Emitting Materials	2		•	
Credit Building Product Disclosure and Optimization	5		•	•
Credit Construction and Demolition Waste Diversion	2	•		•
Indoor Environmental Quality	**13**			
Prereq Construction Management Plan	Required		•	
Prereq Fundamental Air Quality	Required		•	
Prereq No Smoking or Vehicle Idling	Required		•	
Credit Enhanced Air Quality	1		•	
Credit Occupant Experience	7		•	
Credit Accessibility and Inclusion	1		•	
Credit Resilient Spaces	2		•	
Credit Air Quality Testing and Monitoring	2		•	
Project Priorities & Innovation	**10**			
Credit Project Priorities	9			
Credit LEED Accredited Professional	1			
Total	**Possible Points: 110**			

LEED v4.1 BD+C
Project Checklist

Project Name:
Date:

Y ? N

		Credit	Integrative Process		1

			Location and Transportation		16
0	0	0			
		Credit	LEED for Neighborhood Development Location		16
		Credit	Sensitive Land Protection		1
		Credit	High Priority Site and Equitable Development		2
		Credit	Surrounding Density and Diverse Uses		5
		Credit	Access to Quality Transit		5
		Credit	Bicycle Facilities		1
		Credit	Reduced Parking Footprint		1
		Credit	Electric Vehicles		1

			Sustainable Sites		10
0	0	0			
Y		Prereq	Construction Activity Pollution Prevention		Required
		Credit	Site Assessment		1
		Credit	Protect or Restore Habitat		2
		Credit	Open Space		1
		Credit	Rainwater Management		3
		Credit	Heat Island Reduction		2
		Credit	Light Pollution Reduction		1

			Water Efficiency		11
0	0	0			
Y		Prereq	Outdoor Water Use Reduction		Required
Y		Prereq	Indoor Water Use Reduction		Required
Y		Prereq	Building-Level Water Metering		Required
		Credit	Outdoor Water Use Reduction		2
		Credit	Indoor Water Use Reduction		6
		Credit	Optimize Process Water Use		2
		Credit	Water Metering		1

			Energy and Atmosphere		33
0	0	0			
Y		Prereq	Fundamental Commissioning and Verification		Required
Y		Prereq	Minimum Energy Performance		Required
Y		Prereq	Building-Level Energy Metering		Required
Y		Prereq	Fundamental Refrigerant Management		Required
		Credit	Enhanced Commissioning		6
		Credit	Optimize Energy Performance		18
		Credit	Advanced Energy Metering		1
		Credit	Grid Harmonization		2
		Credit	Renewable Energy		5
		Credit	Enhanced Refrigerant Management		1

			Materials and Resources		13
0	0	0			
Y		Prereq	Storage and Collection of Recyclables		Required
		Credit	Building Life-Cycle Impact Reduction		5
		Credit	Environmental Product Declarations		2
		Credit	Sourcing of Raw Materials		2
		Credit	Material Ingredients		2
		Credit	Construction and Demolition Waste Management		2

			Indoor Environmental Quality		16
0	0	0			
Y		Prereq	Minimum Indoor Air Quality Performance		Required
Y		Prereq	Environmental Tobacco Smoke Control		Required
		Credit	Enhanced Indoor Air Quality Strategies		2
		Credit	Low-Emitting Materials		3
		Credit	Construction Indoor Air Quality Management Plan		1
		Credit	Indoor Air Quality Assessment		2
		Credit	Thermal Comfort		1
		Credit	Interior Lighting		2
		Credit	Daylight		3
		Credit	Quality Views		1
		Credit	Acoustic Performance		1

			Innovation		6
0	0	0			
		Credit	Innovation		5
		Credit	LEED Accredited Professional		1

			Regional Priority		4
0	0	0			
		Credit	Regional Priority: Specific Credit		1
		Credit	Regional Priority: Specific Credit		1
		Credit	Regional Priority: Specific Credit		1
		Credit	Regional Priority: Specific Credit		1

			TOTALS	Possible Points:	110
0	0	0			

Certified: 40 to 49 points,　Silver: 50 to 59 points,　Gold: 60 to 79 points,　Platinum 80

LEED v4 BD+C：新建建築與重大改造 (New Construction and Major Renovation)
項目得分表

項目名稱：
日期：

滿足　?　不滿足

	整合過程	1

選址與交通　16
先決條件	LEED 社區開發選址	16
得分點	敏感土地保護	1
得分點	高優先場址	2
得分點	周邊密度和多樣化土地使用	5
得分點	優良公共交通連接	5
得分點	自行車設施	1
得分點	停車面積最低化	1
得分點	綠色機動車	1

可持續場址　10
先決條件	施工污染防治	
得分點	場址評估	1
得分點	場址開發 - 保護和恢復棲息地	2
得分點	空地	1
得分點	雨水管理	3
得分點	降低熱島效應	2
得分點	降低光污染	1

用水效率　11
先決條件	室外用水減量	必要項
先決條件	室內用水減量	必要項
先決條件	建築整體用水計量	
得分點	室外用水減量	2
得分點	室內用水減量	6
得分點	冷卻塔用水	2
得分點	用水計量	1

能源與大氣　33
先決條件	基本調試和查證	必要項
先決條件	最低能源表現	必要項
先決條件	建築整體能源計量	必要項
先決條件	基礎冷媒管理	
得分點	增強調試	6
得分點	能源效率優化	18
得分點	高階能源計量	2
得分點	能源需求反應	3
得分點	可再生能源生產	1
得分點	增強冷媒管理	2
得分點	綠色電力和碳補償	

材料與資源　13
先決條件	可回收物存儲和收集	必要項
先決條件	營建和拆建廢物管理計畫	必要項
得分點	減少建築生命週期中的影響	5
得分點	建築產品分析揭露和優化 - 產品環境要素	2
得分點	聲明	
得分點	建築產品分析揭露和優化 - 原材料的來源和採購	2
得分點	建築產品分析揭露和優化 - 材料成分	2

室內環境品質　16
先決條件	最低室內空氣品質表現	必要項
先決條件	環境煙控	必要項
得分點	增強室內空氣品質策略	2
得分點	低逸散材料	3
得分點	施工期室內空氣品質管制計畫	1
得分點	室內空氣品質評估	2
得分點	熱舒適	1
得分點	室內照明	2
得分點	自然採光	3
得分點	優質視野	1
得分點	聲環境表現	1

創新　6
得分點	創新	5
得分點	LEED Accredited Professional	1

地域優先　4
得分點	地域優先：具體得分點	1
得分點	地域優先：具體得分點	1
得分點	地域優先：具體得分點	1
得分點	地域優先：具體得分點	1

總計　110
認證級：40 至 49 分，銀級：50 至 59 分，金級：60 至 79 分，鉑金級：80 至 110 分
可得分數：80 至 110 分

LEED v4.1 BD+C: Core and Shell
Project Checklist

Project Name:
Date:

Y	?	N			
			Credit	Integrative Process	1

Location and Transportation — 20

			Credit	LEED for Neighborhood Development Location	20
			Credit	Sensitive Land Protection	2
			Credit	High Priority Site and Equitable Development	3
			Credit	Surrounding Density and Diverse Uses	6
			Credit	Access to Quality Transit	6
			Credit	Bicycle Facilities	1
			Credit	Reduced Parking Footprint	1
			Credit	Electric Vehicles	1

Sustainable Sites — 11

Y			Prereq	Construction Activity Pollution Prevention	Required
			Credit	Site Assessment	1
			Credit	Protect or Restore Habitat	2
			Credit	Open Space	1
			Credit	Rainwater Management	3
			Credit	Heat Island Reduction	2
			Credit	Light Pollution Reduction	1
			Credit	Tenant Design and Construction Guidelines	1

Water Efficiency — 11

Y			Prereq	Outdoor Water Use Reduction	Required
Y			Prereq	Indoor Water Use Reduction	Required
Y			Prereq	Building-Level Water Metering	Required
			Credit	Outdoor Water Use Reduction	3
			Credit	Indoor Water Use Reduction	4
			Credit	Optimize Process Water Use	1
			Credit	Water Metering	1

Energy and Atmosphere — 33

Y			Prereq	Fundamental Commissioning and Verification	Required
Y			Prereq	Minimum Energy Performance	Required
Y			Prereq	Building-Level Energy Metering	Required
Y			Prereq	Fundamental Refrigerant Management	Required
			Credit	Enhanced Commissioning	6
			Credit	Optimize Energy Performance	18
			Credit	Advanced Energy Metering	1
			Credit	Grid Harmonization	2
			Credit	Renewable Energy	5
			Credit	Enhanced Refrigerant Management	1

Materials and Resources — 14

Y			Prereq	Storage and Collection of Recyclables	Required
			Credit	Building Life-Cycle Impact Reduction	6
			Credit	Environmental Product Declarations	2
			Credit	Sourcing of Raw Materials	2
			Credit	Material Ingredients	2
			Credit	Construction and Demolition Waste Management	2

Indoor Environmental Quality — 10

Y			Prereq	Minimum Indoor Air Quality Performance	Required
Y			Prereq	Environmental Tobacco Smoke Control	Required
			Credit	Enhanced Indoor Air Quality Strategies	2
			Credit	Low-Emitting Materials	3
			Credit	Construction Indoor Air Quality Management Plan	1
			Credit	Daylight	3
			Credit	Quality Views	1

Innovation — 6

			Credit	Innovation	5
			Credit	LEED Accredited Professional	1

Regional Priority — 4

			Credit	Regional Priority: Specific Credit	1
			Credit	Regional Priority: Specific Credit	1
			Credit	Regional Priority: Specific Credit	1
			Credit	Regional Priority: Specific Credit	1

0	0	0	**TOTALS**	Possible Points:	**110**

Certified: 40 to 49 points, Silver: 50 to 59 points, Gold: 60 to 79 points, Platinum 80 to 110

項目名稱：

日期：

LEED v4 BD+C：核心與外殼 (Core and Shell)
項目得分表

滿足　？　不滿足

滿足	？	不滿足		分數
0	0	0	整合過程	**1**

			選址與交通	**20**
			得分點　LEED 社區開發選址	20
			得分點　敏感土地保護	2
			得分點　高優先地址	3
			得分點　周邊密度和多樣化土地使用	6
			得分點　優良公共交通連接	6
			得分點　自行車設施	1
			得分點　停車面積減量	1
			得分點　綠色機動率	1

滿足			**可持續場址**	**11**
			先決條件　施工污染防治	必要項
			得分點　場址評估	1
			得分點　場址開發 - 保護和恢復棲息地	2
			得分點　空地	1
			得分點　雨水管理	3
			得分點　降低熱島效應	2
			得分點　降低光污染	1
			得分點　租戶設計和施工指南	1

滿足			**用水效率**	**11**
滿足			先決條件　室外用水減量	必要項
滿足			先決條件　室內用水減量	必要項
			先決條件　建築整體用水計量	必要項
			得分點　室外用水減量	2
			得分點　室內用水減量	6
			得分點　冷卻塔用水	2
			得分點　用水計量	1

滿足			**能源與大氣**	**33**
滿足			先決條件　基本調試和查證	必要項
滿足			先決條件　最低能源表現	必要項
滿足			先決條件　建築整體能源計量	必要項
			先決條件　基礎冷媒管理	必要項
			得分點　增強調試	6
			得分點　能源效率優化	18
			得分點　高階能源計量	1
			得分點　能源需求反應	2
			得分點　可再生能源生產	3
			得分點　增強冷媒管理	1
			得分點　綠色電力和碳補償	2

0	0	0	**材料與資源**	**14**
滿足			先決條件　可回收物存儲和收集	必要項
滿足			先決條件　營建拆除建築廢物管理計畫	必要項
			得分點　減少建築生命週期中的影響	6
			得分點　建築產品分析公示和優化 - 產品環境要素聲明	2
			得分點　建築產品分析公示和優化 - 原材料的來源和採購	2
			得分點　建築產品分析公示和優化 - 材料成分	2
			得分點　營建和拆除建築廢物管理	2

0	0	0	**室內環境品質**	**10**
滿足			先決條件　最低室內空氣品質表現	必要項
滿足			先決條件　環境煙控	必要項
			得分點　增強室內空氣品質策略	2
			得分點　低逸散材料	3
			得分點　施工期室內空氣品質管制計畫	1
			得分點　自然採光	3
			得分點　優良視野	1

0	0	0	**創新**	**6**
			得分點　創新	5
			得分點　LEED Accredited Professional	1

0	0	0	**地域優先**	**4**
			得分點　地域優先：具體得分點	1
			得分點　地域優先：具體得分點	1
			得分點　地域優先：具體得分點	1
			得分點　地域優先：具體得分點	1

0	0	0	**總計**	**110**

認證級：40 至 49 分，銀級：50 至 59 分，金級：60 至 79 分，鉑金級：80 至 110 分

可獲分數：110

LEED v4.1 BD+C: Schools
Project Checklist

Project Name

Date:

Y	?	N				
			Credit	Integrative Process		1

Location and Transportation — 15

Y	?	N			
			Credit	LEED for Neighborhood Development Location	15
			Credit	Sensitive Land Protection	1
			Credit	High Priority Site and Equitable Development	2
			Credit	Surrounding Density and Diverse Uses	5
			Credit	Access to Quality Transit	4
			Credit	Bicycle Facilities	1
			Credit	Reduced Parking Footprint	1
			Credit	Electric Vehicles	1

Sustainable Sites — 12

Y	?	N			
Y			Prereq	Construction Activity Pollution Prevention	Required
Y			Prereq	Environmental Site Assessment	Required
			Credit	Site Assessment	1
			Credit	Protect or Restore Habitat	2
			Credit	Open Space	1
			Credit	Rainwater Management	3
			Credit	Heat Island Reduction	2
			Credit	Light Pollution Reduction	1
			Credit	Site Master Plan	1
			Credit	Joint Use of Facilities	1

Water Efficiency — 12

Y	?	N			
Y			Prereq	Outdoor Water Use Reduction	Required
Y			Prereq	Indoor Water Use Reduction	Required
Y			Prereq	Building-Level Water Metering	Required
			Credit	Outdoor Water Use Reduction	2
			Credit	Indoor Water Use Reduction	7
			Credit	Optimize Process Water Use	2
			Credit	Water Metering	1

Energy and Atmosphere — 31

Y	?	N			
Y			Prereq	Fundamental Commissioning and Verification	Required
Y			Prereq	Minimum Energy Performance	Required
Y			Prereq	Building-Level Energy Metering	Required
Y			Prereq	Fundamental Refrigerant Management	Required
			Credit	Enhanced Commissioning	6
			Credit	Optimize Energy Performance	16
			Credit	Advanced Energy Metering	1
			Credit	Grid Harmonization	2
			Credit	Renewable Energy	5
			Credit	Enhanced Refrigerant Management	1

Materials and Resources — 13

Y	?	N			
Y			Prereq	Storage and Collection of Recyclables	Required
			Credit	Building Life-Cycle Impact Reduction	5
			Credit	Environmental Product Declarations	2
			Credit	Sourcing of Raw Materials	2
			Credit	Material Ingredients	2
			Credit	Construction and Demolition Waste Management	2

Indoor Environmental Quality — 16

Y	?	N			
Y			Prereq	Minimum Indoor Air Quality Performance	Required
Y			Prereq	Environmental Tobacco Smoke Control	Required
Y			Prereq	Minimum Acoustic Performance	Required
			Credit	Enhanced Indoor Air Quality Strategies	2
			Credit	Low-Emitting Materials	3
			Credit	Construction Indoor Air Quality Management Plan	1
			Credit	Indoor Air Quality Assessment	2
			Credit	Thermal Comfort	1
			Credit	Interior Lighting	2
			Credit	Daylight	3
			Credit	Quality Views	1
			Credit	Acoustic Performance	1

Innovation — 6

Y	?	N			
			Credit	Innovation	5
			Credit	LEED Accredited Professional	1

Regional Priority — 4

Y	?	N			
			Credit	Regional Priority: Specific Credit	1
			Credit	Regional Priority: Specific Credit	1
			Credit	Regional Priority: Specific Credit	1
			Credit	Regional Priority: Specific Credit	1

TOTALS

				Possible Points:	110
0	0	0	TOTALS		110

Certified: 40 to 49 points, Silver: 50 to 59 points, Gold: 60 to 79 points, Platinum 80 to 110

LEED v4 BD+C: 學校 (Schools)
項目得分表

項目名稱：
日期：

滿足　?　不滿足

滿足	?	不滿足	項目	分數
0	0	0	**整合過程**	**1**
			整合過程	1

滿足	?	不滿足	**選址與交通**	**15**
			先決條件　LEED 社區開發選址	15
			得分點　敏感土地保護	1
			得分點　高優先場址	2
			得分點　周邊密度和多樣化土地使用	5
			得分點　優良公共交通連接	4
			得分點　自行車設施	1
			得分點　停車面積減量	1
			得分點　綠色機動車	1

滿足	?	不滿足	**可持續場址**	**12**
			先決條件　施工污染防治	必要項
			得分點　場址評估	1
			得分點　場址開發 - 保護或恢復棲息地	2
			得分點　空地	1
			得分點　雨水管理	3
			得分點　降低熱島效應	2
			得分點　降低光污染	1
			得分點　場址總圖	1
			得分點　設施共用	1

滿足	?	不滿足	**用水效率**	**12**
			先決條件　室外用水減量	必要項
			先決條件　室內用水減量	必要項
			先決條件　建築整體用水計量	必要項
			得分點　室外用水減量	2
			得分點　室內用水減量	7
			得分點　冷卻塔用水	2
			得分點　用水計量	1

滿足	?	不滿足	**能源與大氣**	**31**
			先決條件　基本調試和查證	必要項
			先決條件　最低能源表現	必要項
			先決條件　建築整體能源計量	必要項
			先決條件　基礎冷媒管理	必要項
			得分點　增強調試	6
			得分點　能源效率優化	16
			得分點　高階能源計量	1
			得分點　能源需求反應	2
			得分點　可再生能源生產	3
			得分點　增強冷媒管理	1
			得分點　綠色電力和碳補償	2

滿足	?	不滿足	**材料與資源**	**13**
			先決條件　可回收物存儲和收集	必要項
			先決條件　營建和拆建廢棄物管理計畫	必要項
			得分點　減少建築生命週期中的影響	5
			得分點　建築產品分析和優化 - 產品環境要素聲明	2
			得分點　建築產品分析和優化 - 原材料的來源和採購	2
			得分點　建築產品分析和優化 - 材料成分	2
			得分點　營建和拆建廢棄物管理	2

滿足	?	不滿足	**室內環境品質**	**16**
			先決條件　最低室內空氣品質表現	必要項
			先決條件　環境煙控	必要項
			先決條件　聲學效果最低品質	必要項
			得分點　增強室內空氣品質策略	2
			得分點　低逸散材料	3
			得分點　施工期室內空氣品質管制計畫	1
			得分點　室內空氣品質評估	2
			得分點　熱舒適	1
			得分點　室內照明	2
			得分點　自然採光	3
			得分點　優良視野	1
			得分點　聲環境表現	1

滿足	?	不滿足	**創新**	**6**
			得分點　創新	5
			得分點　LEED Accredited Professional	1

滿足	?	不滿足	**地域優先**	**4**
			得分點　地域優先：具體得分	1
			得分點　地域優先：具體得分	1
			得分點　地域優先：具體得分	1
			得分點　地域優先：具體得分	1

滿足	?	不滿足	**總計**	可獲分數：**110**
0	0	0		

認證級：40 至 49 分，銀級：50 至 59 分，金級：60 至 79 分，鉑金級：80 至 110 分

LEED v4.1 BD+C: Retail
Project Checklist

Project Name:
Date:

Y	?	N			
			Credit	Integrative Process	1

Y	?	N		Location and Transportation	16
			Credit	LEED for Neighborhood Development Location	16
			Credit	Sensitive Land Protection	1
			Credit	High Priority Site and Equitable Development	2
			Credit	Surrounding Density and Diverse Uses	5
			Credit	Access to Quality Transit	5
			Credit	Bicycle Facilities	1
			Credit	Reduced Parking Footprint	1
			Credit	Electric Vehicles	1

Y	?	N		Sustainable Sites	10
Y			Prereq	Construction Activity Pollution Prevention	Required
			Credit	Site Assessment	1
			Credit	Protect or Restore Habitat	2
			Credit	Open Space	1
			Credit	Rainwater Management	3
			Credit	Heat Island Reduction	2
			Credit	Light Pollution Reduction	1

Y	?	N		Water Efficiency	12
Y			Prereq	Outdoor Water Use Reduction	Required
Y			Prereq	Indoor Water Use Reduction	Required
Y			Prereq	Building-Level Water Metering	Required
			Credit	Outdoor Water Use Reduction	2
			Credit	Indoor Water Use Reduction	7
			Credit	Optimize Process Water Use	2
			Credit	Water Metering	1

Y	?	N		Energy and Atmosphere	33
Y			Prereq	Fundamental Commissioning and Verification	Required
Y			Prereq	Minimum Energy Performance	Required
Y			Prereq	Building-Level Energy Metering	Required
Y			Prereq	Fundamental Refrigerant Management	Required
			Credit	Enhanced Commissioning	6
			Credit	Optimize Energy Performance	18
			Credit	Advanced Energy Metering	1
			Credit	Grid Harmonization	2
			Credit	Renewable Energy	5
			Credit	Enhanced Refrigerant Management	1

Y	?	N		Materials and Resources	13
Y			Prereq	Storage and Collection of Recyclables	Required
			Credit	Building Life-Cycle Impact Reduction	5
			Credit	Environmental Product Declarations	2
			Credit	Sourcing of Raw Materials	2
			Credit	Material Ingredients	2
			Credit	Construction and Demolition Waste Management	2

Y	?	N		Indoor Environmental Quality	15
Y			Prereq	Minimum Indoor Air Quality Performance	Required
Y			Prereq	Environmental Tobacco Smoke Control	Required
			Credit	Enhanced Indoor Air Quality Strategies	2
			Credit	Low-Emitting Materials	3
			Credit	Construction Indoor Air Quality Management Plan	1
			Credit	Indoor Air Quality Assessment	2
			Credit	Thermal Comfort	1
			Credit	Interior Lighting	2
			Credit	Daylight	3
			Credit	Quality Views	1

Y	?	N		Innovation	6
			Credit	Innovation	5
			Credit	LEED Accredited Professional	1

Y	?	N		Regional Priority	4
			Credit	Regional Priority: Specific Credit	1
			Credit	Regional Priority: Specific Credit	1
			Credit	Regional Priority: Specific Credit	1
			Credit	Regional Priority: Specific Credit	1

0	0	0	TOTALS	Possible Points:	110

Certified: 40 to 49 points, Silver: 50 to 59 points, Gold: 60 to 79 points, Platinum 80 to 110

LEED v4 BD+C: 零售 (Retail)
項目得分表

項目名稱：
日期：

滿足	?	不滿足			
			得分點	整合過程	1

滿足	?	不滿足		選址與交通	16
0	0	0			
			得分點	LEED 社區開發選址	16
			得分點	敏感土地保護	1
			得分點	高優先場址	2
			得分點	周邊密度和多樣化土地使用	5
			得分點	優良公共交通連接	5
			得分點	自行車設施	1
			得分點	停車面積減量	1
			得分點	綠色機動車	1

滿足	?	不滿足		可持續績址	10
0	0	0			
滿足			先決條件	施工污染防治	必要項
			得分點	場址評估	1
			得分點	場址開發－保護和恢復棲息地	2
			得分點	空地	1
			得分點	雨水管理	3
			得分點	降低熱島效應	2
			得分點	降低光污染	1

滿足	?	不滿足		用水效率	12
0	0	0			
滿足			先決條件	室外用水減量	必要項
滿足			先決條件	室內用水減量	必要項
滿足			先決條件	建築整體用水計量	必要項
			得分點	室外用水減量	2
			得分點	室內用水減量	7
			得分點	冷卻塔用水	2
			得分點	用水計量	1

滿足	?	不滿足		能源與大氣	33
0	0	0			
滿足			先決條件	基本調試和查證	必要項
滿足			先決條件	最低能源表現	必要項
滿足			先決條件	建築整體能源計量	必要項
滿足			先決條件	基礎冷媒管理	必要項
			得分點	增強調試	6
			得分點	能源效率優化	18
			得分點	高階能源計量	1
			得分點	需求反應	2
			得分點	可再生能源生產	3
			得分點	增強冷媒管理	1
			得分點	綠色電力和碳補償	2

滿足	?	不滿足		材料與資源	13
0	0	0			
滿足			先決條件	可回收物存儲和收集	必要項
滿足			先決條件	營建和拆建廢棄物管理計畫	必要項
			得分點	減少建築生命週期中的影響	5
			得分點	建築產品分析公示和優化 - 產品環境要素聲明	2
			得分點	建築產品分析公示和優化 - 原材料的來源和採購	2
			得分點	建築產品分析公示和優化 - 材料成分	2
			得分點	營建和拆建廢棄物管理	2

滿足	?	不滿足		室內環境品質	15
0	0	0			
滿足			先決條件	最低室內空氣品質表現	必要項
滿足			先決條件	環境煙控	必要項
			得分點	增強室內空氣品質策略	2
			得分點	低逸散材料	3
			得分點	施工期室內空氣品質管制計畫	1
			得分點	室內空氣品質評估	2
			得分點	熱舒適	1
			得分點	室內照明	2
			得分點	自然採光	3
			得分點	優良視野	1

滿足	?	不滿足		創新	6
0	0	0			
			得分點	創新	5
			得分點	LEED Accredited Professional	1

滿足	?	不滿足		地域優先	4
0	0	0			
			得分點	地域優先：具體得分點	1
			得分點	地域優先：具體得分點	1
			得分點	地域優先：具體得分點	1
			得分點	地域優先：具體得分點	1

滿足	?	不滿足		總計	110
0	0	0			

認證級：40 至 49 分，銀級：50 至 59 分，金級：60 至 79 分，鉑金級：80 至 110 分

可得分數：110

LEED v4.1 BD+C: Data Centers
Project Checklist

Project Name:
Date:

Y ? N

			Credit	Integrative Process	**1**

0	0	0	**Location and Transportation**		**16**
			Credit	LEED for Neighborhood Development Location	16
			Credit	Sensitive Land Protection	1
			Credit	High Priority Site and Equitable Development	2
			Credit	Surrounding Density and Diverse Uses	5
			Credit	Access to Quality Transit	5
			Credit	Bicycle Facilities	1
			Credit	Reduced Parking Footprint	1
			Credit	Electric Vehicles	1

0	0	0	**Sustainable Sites**		**10**
Y			Prereq	Construction Activity Pollution Prevention	Required
			Credit	Site Assessment	1
			Credit	Protect or Restore Habitat	2
			Credit	Open Space	1
			Credit	Rainwater Management	3
			Credit	Heat Island Reduction	2
			Credit	Light Pollution Reduction	1

0	0	0	**Water Efficiency**		**11**
Y			Prereq	Outdoor Water Use Reduction	Required
Y			Prereq	Indoor Water Use Reduction	Required
Y			Prereq	Building-Level Water Metering	Required
			Credit	Outdoor Water Use Reduction	2
			Credit	Indoor Water Use Reduction	6
			Credit	Optimize Process Water Use	2
			Credit	Water Metering	1

0	0	0	**Energy and Atmosphere**		**33**
Y			Prereq	Fundamental Commissioning and Verification	Required
Y			Prereq	Minimum Energy Performance	Required
Y			Prereq	Building-Level Energy Metering	Required
Y			Prereq	Fundamental Refrigerant Management	Required
			Credit	Enhanced Commissioning	6
			Credit	Optimize Energy Performance	18
			Credit	Advanced Energy Metering	1
			Credit	Grid Harmonization	2
			Credit	Renewable Energy	5
			Credit	Enhanced Refrigerant Management	1

0	0	0	**Materials and Resources**		**13**
Y			Prereq	Storage and Collection of Recyclables	Required
			Credit	Building Life-Cycle Impact Reduction	5
			Credit	Environmental Product Declarations	2
			Credit	Sourcing of Raw Materials	2
			Credit	Material Ingredients	2
			Credit	Construction and Demolition Waste Management	2

0	0	0	**Indoor Environmental Quality**		**16**
Y			Prereq	Minimum Indoor Air Quality Performance	Required
Y			Prereq	Environmental Tobacco Smoke Control	Required
			Credit	Enhanced Indoor Air Quality Strategies	2
			Credit	Low-Emitting Materials	3
			Credit	Construction Indoor Air Quality Management Plan	1
			Credit	Indoor Air Quality Assessment	2
			Credit	Thermal Comfort	1
			Credit	Interior Lighting	2
			Credit	Daylight	3
			Credit	Quality Views	1
			Credit	Acoustic Performance	1

0	0	0	**Innovation**		**6**
			Credit	Innovation	5
			Credit	LEED Accredited Professional	1

0	0	0	**Regional Priority**		**4**
			Credit	Regional Priority: Specific Credit	1
			Credit	Regional Priority: Specific Credit	1
			Credit	Regional Priority: Specific Credit	1
			Credit	Regional Priority: Specific Credit	1

0	0	0	**TOTALS**	Possible Points:	**110**

Certified: 40 to 49 points, Silver: 50 to 59 points, Gold: 60 to 79 points, Platinum 80 to 110

LEED v4 BD+C: 資料中心 (Data Centers)
項目得分表

項目名稱：
日期：

滿足　?　不滿足

滿足	?	不滿足		
			整合過程	1

滿足	?	不滿足	選址與交通	16
0	0	0		
			先決條件　LEED 社區開發選址	必要項
			得分點　敏感土地保護	1
			得分點　高優先地址	2
			得分點　周邊密度和多樣化土地使用	5
			得分點　優良公共交通連接	5
			得分點　自行車設施	1
			得分點　停車面積減量	1
			得分點　綠色機動車	1

滿足	?	不滿足	可持續場址	10
0	0	0		
			先決條件　施工污染防治	必要項
			得分點　場址評估	1
			得分點　場址開發 - 保護和恢復棲息地	2
			得分點　空地	1
			得分點　雨水管理	3
			得分點　降低熱島效應	2
			得分點　降低光污染	1

滿足	?	不滿足	用水效率	11
0	0	0		
			先決條件　室外用水減量	必要項
			先決條件　室內用水減量	必要項
			先決條件　建築整體用水計量	必要項
			得分點　室外用水減量	2
			得分點　室內用水減量	6
			得分點　冷卻塔用水	2
			得分點　用水計量	1

滿足	?	不滿足	能源與大氣	33
0	0	0		
			先決條件　基本調試和查證	必要項
			先決條件　最低能源表現	必要項
			先決條件　建築整體能源計量	必要項
			先決條件　基礎冷媒管理	必要項
			得分點　增強調試	6
			得分點　能源效率優化	18
			得分點　高階能源計量	1
			得分點　能源需求反應	2
			得分點　可再生能源生產	3
			得分點　增強冷媒管理	1
			得分點　綠色電力和碳補償	2

滿足	?	不滿足	材料與資源	13
0	0	0		
			先決條件　可回收物存儲和收集	必要項
			先決條件　營建和拆除廢物管理計畫	必要項
			得分點　減少建築生命週期中的影響	5
			得分點　建築產品分析公示和優化 - 產品環境要素聲明	2
			得分點　建築產品分析公示和優化 - 原材料的來源和採購	2
			得分點　建築產品分析公示和優化 - 材料成分	2
			得分點　營建和拆除廢物管理	2

滿足	?	不滿足	室內環境品質	16
0	0	0		
			先決條件　最低室內空氣品質表現	必要項
			先決條件　環境煙控	必要項
			得分點　增強室內空氣品質策略	2
			得分點　低逸散材料	3
			得分點　施工期室內空氣品質管制計畫	1
			得分點　室內空氣品質評估	2
			得分點　熱舒適	1
			得分點　室內照明	2
			得分點　自然採光	3
			得分點　優良視野	1
			得分點　聲學環境表現	1

滿足	?	不滿足	創新	6
0	0	0		
			得分點　創新	5
			得分點　LEED Accredited Professional	1

滿足	?	不滿足	地域優先	4
0	0	0		
			得分點　地域優先：具體得分點	1
			得分點　地域優先：具體得分點	1
			得分點　地域優先：具體得分點	1
			得分點　地域優先：具體得分點	1

			總計	110
0	0	0		可獲分數：110

認證級：40 至 49 分，銀級：50 至 59 分，金級：60 至 79 分，鉑金級：80 至 110 分

LEED v4.1 BD+C: Warehouses and Distribution Centers

Project Checklist

Project Name:

Date:

Y	?	N			
			Credit	Integrative Process	1

0	0	0	**Location and Transportation**		16
			Credit	LEED for Neighborhood Development Location	16
			Credit	Sensitive Land Protection	1
			Credit	High Priority Site and Equitable Development	2
			Credit	Surrounding Density and Diverse Uses	5
			Credit	Access to Quality Transit	5
			Credit	Bicycle Facilities	1
			Credit	Reduced Parking Footprint	1
			Credit	Electric Vehicles	

0	0	0	**Sustainable Sites**		10
Y			Prereq	Construction Activity Pollution Prevention	Required
			Credit	Site Assessment	1
			Credit	Protect or Restore Habitat	2
			Credit	Open Space	1
			Credit	Rainwater Management	3
			Credit	Heat Island Reduction	2
			Credit	Light Pollution Reduction	1

0	0	0	**Water Efficiency**		11
Y			Prereq	Outdoor Water Use Reduction	Required
Y			Prereq	Indoor Water Use Reduction	Required
Y			Prereq	Building-Level Water Metering	Required
			Credit	Outdoor Water Use Reduction	2
			Credit	Indoor Water Use Reduction	6
			Credit	Optimize Process Water Use	2
			Credit	Water Metering	1

0	0	0	**Energy and Atmosphere**		33
Y			Prereq	Fundamental Commissioning and Verification	Required
Y			Prereq	Minimum Energy Performance	Required
Y			Prereq	Building-Level Energy Metering	Required
Y			Prereq	Fundamental Refrigerant Management	Required
			Credit	Enhanced Commissioning	6
			Credit	Optimize Energy Performance	18
			Credit	Advanced Energy Metering	1
			Credit	Grid Harmonization	2
			Credit	Renewable Energy	5
			Credit	Enhanced Refrigerant Management	1

0	0	0	**Materials and Resources**		13
Y			Prereq	Storage and Collection of Recyclables	Required
			Credit	Building Life-Cycle Impact Reduction	5
			Credit	Environmental Product Declarations	2
			Credit	Sourcing of Raw Materials	2
			Credit	Material Ingredients	2
			Credit	Construction and Demolition Waste Management	2

0	0	0	**Indoor Environmental Quality**		16
Y			Prereq	Minimum Indoor Air Quality Performance	Required
Y			Prereq	Environmental Tobacco Smoke Control	Required
			Credit	Enhanced Indoor Air Quality Strategies	2
			Credit	Low-Emitting Materials	3
			Credit	Construction Indoor Air Quality Management Plan	1
			Credit	Indoor Air Quality Assessment	2
			Credit	Thermal Comfort	1
			Credit	Interior Lighting	2
			Credit	Daylight	3
			Credit	Quality Views	1
			Credit	Acoustic Performance	1

0	0	0	**Innovation**		6
			Credit	Innovation	5
			Credit	LEED Accredited Professional	1

0	0	0	**Regional Priority**		4
			Credit	Regional Priority: Specific Credit	1
			Credit	Regional Priority: Specific Credit	1
			Credit	Regional Priority: Specific Credit	1
			Credit	Regional Priority: Specific Credit	1

0	0	0	**TOTALS**	Possible Points:	110
				0 to 49 points, **Silver:** 50 to 59 points, **Gold:** 60 to 79 points, **Platinu**	

LEED v4 BD+C: 倉儲和配送中心 (Warehouses and Distribution Centers)

項目得分表

項目名稱：
日期：

滿足　?　不滿足

		整合過程	**1**

選址與交通　16

滿足 0　0

得分點	LEED 社區開發選址	16
得分點	敏感土地保護	1
得分點	高優先場址	2
得分點	周邊密度和多樣化土地使用	5
得分點	優良公共交通連接	5
得分點	自行車設施	1
得分點	停車面積最大量	1
得分點	綠色機動車	1

可持續場址　10

滿足 0　0

先決條件	施工污染防治	滿足
得分點	場址評估	1
得分點	場址開發 - 保護和恢復棲息地	2
得分點	空地	1
得分點	雨水管理	3
得分點	降低熱島效應	2
得分點	降低光污染	1

用水效率　11

滿足 0　0

先決條件	室內用水減量	滿足
先決條件	室外用水減量	滿足
先決條件	建築整體用水計量	滿足
得分點	室外用水減量	2
得分點	室內用水減量	6
得分點	冷卻塔用水	2
得分點	用水計量	1

能源與大氣　33

滿足 0　0

先決條件	基本調試和查證	滿足
先決條件	最低能源表現	滿足
先決條件	建築整體能源計量	滿足
先決條件	基礎冷媒管理	滿足
得分點	增強調試	6
得分點	能源效率優化	18
得分點	高階能源計量	1
得分點	能源需求反應	2
得分點	可再生能源生產	3
得分點	增強冷媒管理	1
得分點	綠色電力和碳補償	2

材料與資源　13

滿足 0　0

先決條件	可回收物存儲和收集	滿足
先決條件	營建和拆建廢物管理計畫	滿足
得分點	減少建築生命週期中的影響	5
得分點	建築產品分析和優化 - 產品環境要素	2
得分點	聲明	
得分點	建築產品分析和優化 - 原材料的來源和採購	2
得分點	建築產品分析和優化 - 材料成分	2
得分點	營建和拆建廢物管理	2

室內環境品質　16

滿足 0　0

先決條件	最低室內空氣品質表現	滿足
先決條件	環境煙控	滿足
得分點	增強室內空氣品質策略	2
得分點	低逸散材料	3
得分點	施工期室內空氣品質管制計畫	1
得分點	室內空氣品質評估	2
得分點	熱舒適	1
得分點	室內照明	2
得分點	自然採光	3
得分點	優良視野	1
得分點	整體聲環境表現	1

創新　6

0　0

得分點	創新	5
得分點	LEED Accredited Professional	1

地域優先　4

0　0

得分點	地域優先：具體得分點	1
得分點	地域優先：具體得分點	1
得分點	地域優先：具體得分點	1
得分點	地域優先：具體得分點	1

總計　110

0　0　0

可優分數：110

認證級：40 至 49 分，銀級：50 至 59 分，金級：60 至 79 分，鉑金級：80 至 110 分

LEED v4.1 BD+C: Hospitality
Project Checklist

Project Name:
Date:

Y ? N

		Credit	Integrative Process	1

			Location and Transportation	16
0	0	0		
		Credit	LEED for Neighborhood Development Location	16
		Credit	Sensitive Land Protection	1
		Credit	High Priority Site and Equitable Development	2
		Credit	Surrounding Density and Diverse Uses	5
		Credit	Access to Quality Transit	5
		Credit	Bicycle Facilities	1
		Credit	Reduced Parking Footprint	1
		Credit	Electric Vehicles	1

			Sustainable Sites	10
0	0	0		
Y			Prereq	Construction Activity Pollution Prevention	Required
			Credit	Site Assessment	1
			Credit	Protect or Restore Habitat	2
			Credit	Open Space	1
			Credit	Rainwater Management	3
			Credit	Heat Island Reduction	2
			Credit	Light Pollution Reduction	1

			Water Efficiency	11
0	0	0		
Y			Prereq	Outdoor Water Use Reduction	Required
Y			Prereq	Indoor Water Use Reduction	Required
Y			Prereq	Building-Level Water Metering	Required
			Credit	Outdoor Water Use Reduction	2
			Credit	Indoor Water Use Reduction	6
			Credit	Optimize Process Water Use	2
			Credit	Water Metering	1

			Energy and Atmosphere	33
0	0	0		
Y			Prereq	Fundamental Commissioning and Verification	Required
Y			Prereq	Minimum Energy Performance	Required
Y			Prereq	Building-Level Energy Metering	Required
Y			Prereq	Fundamental Refrigerant Management	Required
			Credit	Enhanced Commissioning	6
			Credit	Optimize Energy Performance	18
			Credit	Advanced Energy Metering	1
			Credit	Grid Harmonization	2
			Credit	Renewable Energy	5
			Credit	Enhanced Refrigerant Management	1

			Materials and Resources	13
0	0	0		
Y			Prereq	Storage and Collection of Recyclables	Required
			Credit	Building Life-Cycle Impact Reduction	5
			Credit	Environmental Product Declarations	2
			Credit	Sourcing of Raw Materials	2
			Credit	Material Ingredients	2
			Credit	Construction and Demolition Waste Management	2

			Indoor Environmental Quality	16
0	0	0		
Y			Prereq	Minimum Indoor Air Quality Performance	Required
Y			Prereq	Environmental Tobacco Smoke Control	Required
			Credit	Enhanced Indoor Air Quality Strategies	2
			Credit	Low-Emitting Materials	3
			Credit	Construction Indoor Air Quality Management Plan	1
			Credit	Indoor Air Quality Assessment	2
			Credit	Thermal Comfort	1
			Credit	Interior Lighting	2
			Credit	Daylight	3
			Credit	Quality Views	1
			Credit	Acoustic Performance	1

			Innovation	6
0	0	0		
			Credit	Innovation	5
			Credit	LEED Accredited Professional	1

			Regional Priority	4
0	0	0		
			Credit	Regional Priority: Specific Credit	1
			Credit	Regional Priority: Specific Credit	1
			Credit	Regional Priority: Specific Credit	1
			Credit	Regional Priority: Specific Credit	1

			TOTALS	Possible Points:	110
0	0	0			

40 to 49 points, Silver: 50 to 59 points, Gold: 60 to 79 points, Platinum

項目名稱：
日期：

LEED v4 BD+C: 賓館接待 (Hospitality)
項目得分表

滿足　?　不滿足

	滿足	?	不滿足	
整合過程				1

選址與交通 16
	滿足	?	不滿足	
	0	0	0	
先決條件 LEED 社區開發選址				16
得分點 敏感土地保護				1
得分點 高優先場址				2
得分點 周邊密度和多樣化土地使用				5
得分點 優良公共交通連接				5
得分點 自行車設施				1
得分點 停車面積減量				1
得分點 綠色機動車				1

可持續場址 10
	滿足	?	不滿足	
	0	0	0	
先決條件 施工污染防治				必要項
得分點 場址評估				1
得分點 場址開發－保護和恢復棲息地				2
得分點 空地				1
得分點 雨水管理				3
得分點 降低熱島效應				2
得分點 降低光污染				1

用水效率 11
	滿足	?	不滿足	
	0	0	0	
先決條件 室外用水減量				必要項
先決條件 室內用水減量				必要項
先決條件 建築整體用水計量				必要項
得分點 室外用水減量				2
得分點 室內用水減量				6
得分點 冷卻塔用水				2
得分點 用水計量				1

能源與大氣 33
	滿足	?	不滿足	
	0	0	0	
先決條件 基本調試和查證				必要項
先決條件 最低能源表現				必要項
先決條件 建築整體能源計量				必要項
先決條件 基礎冷媒管理				必要項
得分點 增強調試				6
得分點 能源效率優化				18
得分點 高階能源計量				1
得分點 能源需求反應				2
得分點 可再生能源生產				3
得分點 增強冷媒管理				1
得分點 綠色電力和碳補償				2

材料與資源 13
	滿足	?	不滿足	
	0	0	0	
先決條件 可回收物存儲和收集				必要項
先決條件 營建和拆建廢棄物管理計畫				必要項
得分點 減小建築生命週期中的影響				5
得分點 建築產品分析公示和優化－產品環境要素				2
得分點 聲明				
得分點 建築產品分析公示和優化－原材料的來源和採購				2
得分點 建築產品分析公示和優化－材料成分				2
得分點 營建和拆建廢棄物管理				2

室內環境品質 16
	滿足	?	不滿足	
	0	0	0	
先決條件 最低室內空氣品質表現				必要項
先決條件 環境煙控				必要項
得分點 增強室內空氣品質策略				2
得分點 低逸散材料				3
得分點 施工期室內空氣品質管制計畫				1
得分點 室內空氣品質評估				2
得分點 熱舒適				1
得分點 室內照明				2
得分點 自然採光				3
得分點 優良視野				1
得分點 聲環境表現				1

創新 6
	滿足	?	不滿足	
	0	0	0	
創新				5
LEED Accredited Professional				1

地域優先 4
	滿足	?	不滿足	
	0	0	0	
得分點 地域優先：具體得分點				1
得分點 地域優先：具體得分點				1
得分點 地域優先：具體得分點				1
得分點 地域優先：具體得分點				1

總計 110
	滿足	?	不滿足	
	0	0	0	

認證級：40 至 49 分，銀級：50 至 59 分，金級：60 至 79 分，鉑金級：80 至 110 分　　可獲分數：110

LEED v4.1 BD+C: Healthcare
Project Checklist

Project Name:
Date:

Y ? N

		Points
Prereq	Integrative Project Planning and Design	Required
Credit	Integrative Process	1

Location and Transportation — 9

		Points
Credit	LEED for Neighborhood Development Location	9
Credit	Sensitive Land Protection	1
Credit	High Priority Site and Equitable Development	2
Credit	Surrounding Density and Diverse Uses	2
Credit	Access to Quality Transit	1
Credit	Bicycle Facilities	1
Credit	Reduced Parking Footprint	1
Credit	Electric Vehicles	1

Sustainable Sites — 9

		Points
Prereq	Construction Activity Pollution Prevention	Required
Prereq	Environmental Site Assessment	Required
Credit	Site Assessment	1
Credit	Protect or Restore Habitat	1
Credit	Open Space	1
Credit	Rainwater Management	2
Credit	Heat Island Reduction	1
Credit	Light Pollution Reduction	1
Credit	Places of Respite	1
Credit	Direct Exterior Access	1

Water Efficiency — 11

		Points
Prereq	Outdoor Water Use Reduction	Required
Prereq	Indoor Water Use Reduction	Required
Prereq	Building-Level Water Metering	Required
Credit	Outdoor Water Use Reduction	1
Credit	Indoor Water Use Reduction	7
Credit	Optimize Process Water Use	2
Credit	Water Metering	1

Energy and Atmosphere — 35

		Points
Prereq	Fundamental Commissioning and Verification	Required
Prereq	Minimum Energy Performance	Required
Prereq	Building-Level Energy Metering	Required
Prereq	Fundamental Refrigerant Management	Required
Credit	Enhanced Commissioning	6
Credit	Optimize Energy Performance	20
Credit	Advanced Energy Metering	2
Credit	Grid Harmonization	5
Credit	Renewable Energy	5
Credit	Enhanced Refrigerant Management	1

Materials and Resources — 19

		Points
Prereq	Storage and Collection of Recyclables	Required
Prereq	PBT Source Reduction- Mercury	Required
Credit	Building Life-Cycle Impact Reduction	5
Credit	Environmental Product Declarations	2
Credit	Sourcing of Raw Materials	2
Credit	Material Ingredients	2
Credit	PBT Source Reduction- Mercury	1
Credit	PBT Source Reduction- Lead, Cadmium and Copper	2
Credit	Furniture and Medical Furnishings	2
Credit	Design for Flexibility	1
Credit	Construction and Demolition Waste Management	2

Indoor Environmental Quality — 16

		Points
Prereq	Minimum Indoor Air Quality Performance	Required
Prereq	Environmental Tobacco Smoke Control	Required
Credit	Enhanced Indoor Air Quality Strategies	2
Credit	Low-Emitting Materials	3
Credit	Construction Indoor Air Quality Management Plan	1
Credit	Indoor Air Quality Assessment	2
Credit	Thermal Comfort	1
Credit	Interior Lighting	1
Credit	Daylight	2
Credit	Quality Views	2
Credit	Acoustic Performance	2

Innovation — 6

		Points
Credit	Innovation	5
Credit	LEED Accredited Professional	1

Regional Priority — 4

		Points
Credit	Regional Priority: Specific Credit	1
Credit	Regional Priority: Specific Credit	1
Credit	Regional Priority: Specific Credit	1
Credit	Regional Priority: Specific Credit	1

TOTALS — Possible Points: 110

Certified: 40 to 49 points,　Silver: 50 to 59 points,　Gold: 60 to 79 points,　Platinum 80 to 110

LEED v4 BD+C: 醫療保健 (Healthcare)
項目得分表

項目名稱：
日期：

滿足　?　不滿足

	必要項
整合專案計畫與設計	1
整合過程	

選址與交通　9

		必要項
先決條件	LEED 社區開發選址	9
得分點	敏感土地保護	1
得分點	高優先場址	2
得分點	周邊密度和多樣化土地使用	1
得分點	優良公共交通連接	2
得分點	自行車設施	1
得分點	停車面積減量	1
得分點	綠色機動車	1

可持續場址　9

		必要項
先決條件	施工污染防治	1
先決條件	場址環境評估	1
得分點	場址評估	1
得分點	空地	1
得分點	雨水管理	2
得分點	降低熱島效應	1
得分點	降低光污染	1
得分點	身心舒緩場地	1
得分點	直接室外人口	1

用水效率　11

		必要項
先決條件	室外用水減量	1
先決條件	室內用水減量	1
先決條件	建築整體用水計量	1
得分點	室外用水減量	7
得分點	室內用水減量	2
得分點	冷卻塔用水	1
得分點	用水計量	1

能源與大氣　35

		必要項
先決條件	基本調試和查證	1
先決條件	最低能源表現	1
先決條件	建築整體能源計量	1
得分點	增強調試	6
得分點	能源效率優化	20
得分點	高階能源計量	1
得分點	能源需求反應	2
得分點	可再生能源生產	3
得分點	增強冷媒管理	1
得分點	綠色電力和碳補償	2

材料與資源　19

		必要項
先決條件	可回收物存儲和收集	必要項
先決條件	營建和拆除廢棄物管理計畫	必要項
先決條件	PBT 來源減量 - 汞	必要項
得分點	減少建築生命週期中的影響	5
得分點	建築產品分析公示和優化 - 產品環境要素聲明	2
得分點	建築產品分析公示和優化 - 原材料的來源和採購	2
得分點	建築產品分析公示和優化 - 材料成分	2
得分點	PBT 來源減量 - 鉛、鎘和銅	1
得分點	PBT 來源減量 - 汞	1
得分點	傢俱和醫療設備	2
得分點	靈活性設計	1
得分點	營建和拆除廢棄物管理	2

室內環境品質　16

		必要項
先決條件	最低室內空氣品質表現	必要項
先決條件	環境煙控	必要項
得分點	增強室內空氣品質策略	2
得分點	低逸散材料	3
得分點	施工期室內空氣品質管理計畫	1
得分點	室內空氣品質評估	2
得分點	熱舒適	1
得分點	室內照明	1
得分點	自然採光	2
得分點	優良視野	1
得分點	整體聲表現	2

創新　6

得分點	創新	5
得分點	LEED Accredited Professional	1

地域優先　4

得分點	地域優先：具體得分點	1
得分點	地域優先：具體得分點	1
得分點	地域優先：具體得分點	1
得分點	地域優先：具體得分點	1

總計　110

可獲分數：110

認證級：40 至 49 分，銀級：50 至 59 分，金級：60 至 79 分，鉑金級：80 至 110 分

國家圖書館出版品預行編目資料

美國綠建築LEED基礎知識與應用：Green Associate
認證／江軍，林巧文著.--二版.--臺北市：
五南圖書出版股份有限公司, 2025.02
面；　公分
ISBN 978-626-423-041-4(平裝)

1.CST: 綠建築　2.CST: 建築節能

441.577　　　　　　　　　113019273

5G34

美國綠建築LEED基礎知識與應用
—— Green Associate認證

作　者 ― 江　軍（44.4）　林巧文

編輯主編 ― 王正華

責任編輯 ― 張維文

封面設計 ― 姚孝慈

出 版 者 ― 五南圖書出版股份有限公司

發 行 人 ― 楊榮川

總 經 理 ― 楊士清

總 編 輯 ― 楊秀麗

地　　址：106台北市大安區和平東路二段339號4樓

電　　話：(02)2705-5066　　傳　真：(02)2706-6100

網　　址：https://www.wunan.com.tw

電子郵件：wunan@wunan.com.tw

劃撥帳號：01068953

戶　　名：五南圖書出版股份有限公司

法律顧問　林勝安律師

出版日期　2016年11月初版一刷
　　　　　2025年 2 月二版一刷

定　　價　新臺幣700元

※版權所有·欲利用本書內容，必須徵求本公司同意※

五 南
WU-NAN

全新官方臉書

五南讀書趣

WUNAN
Books
since1966

Facebook 按讚

👍 1秒變文青

🔍 五南讀書趣 Wunan Books

★ 專業實用有趣
★ 搶先書籍開箱
★ 獨家優惠好康

不定期舉辦抽獎
贈書活動喔！！！

經典永恆・名著常在

五十週年的獻禮 —— 經典名著文庫

五南，五十年了，半個世紀，人生旅程的一大半，走過來了。

思索著，邁向百年的未來歷程，能為知識界、文化學術界作些什麼？

在速食文化的生態下，有什麼值得讓人雋永品味的？

歷代經典・當今名著，經過時間的洗禮，千錘百鍊，流傳至今，光芒耀人；

不僅使我們能領悟前人的智慧，同時也增深加廣我們思考的深度與視野。

我們決心投入巨資，有計畫的系統梳選，成立「經典名著文庫」，

希望收入古今中外思想性的、充滿睿智與獨見的經典、名著。

這是一項理想性的、永續性的巨大出版工程。

不在意讀者的眾寡，只考慮它的學術價值，力求完整展現先哲思想的軌跡；

為知識界開啟一片智慧之窗，營造一座百花綻放的世界文明公園，

任君遨遊、取菁吸蜜、嘉惠學子！